We Are All Whalers

We Are All Whalers

The Plight of Whales and Our Responsibility

Michael J. Moore

The University of Chicago Press

Chicago and London

The University of Chicago Press, Chicago 60637
The University of Chicago Press, Ltd., London
© 2021 by Woods Hole Oceanographic Institution
Published 2021
Printed in the United States of America

30 29 28 27 26 25 24 23 22 21   1 2 3 4 5

ISBN-13: 978-0-226-80304-3 (cloth)
ISBN-13: 978-0-226-80318-0 (e-book)
DOI: https://doi.org/10.7208/chicago/9780226803180.001.0001

Library of Congress Cataloging-in-Publication Data

Names: Moore, Michael J. (Marine biologist), author.
Title: We are all whalers : the plight of whales and our responsibility /
    Michael J. Moore.
Description: Chicago ; London : The University of Chicago Press, 2021. |
    Includes bibliographical references and index.
Identifiers: LCCN 2021003573 | ISBN 9780226803043 (cloth) |
    ISBN 9780226803180 (ebook)
Subjects: LCSH: Northern right whale—Effect of human beings on. |
    Bowhead whale—Effect of human beings on. | Right whales—
    Research.
Classification: LCC QL737.C423 M667 2021 | DDC 599.5/273—dc23
LC record available at https://lccn.loc.gov/2021003573

♾ This paper meets the requirements of ANSI/NISO Z39.48-1992
(Permanence of Paper).

Contents

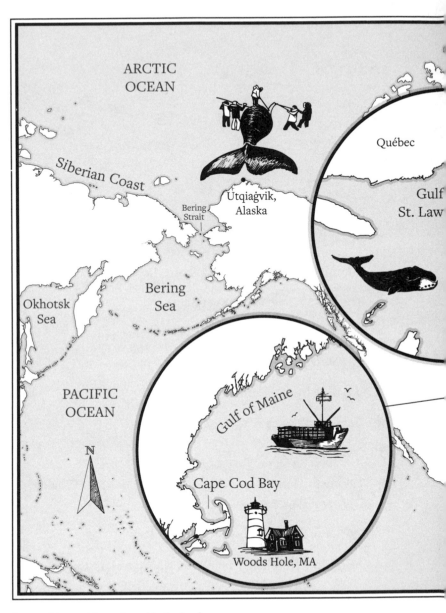

Areas of the world mentioned in the book.
© Woods Hole Oceanographic Institution, Natalie Renier, WHOI Creative.

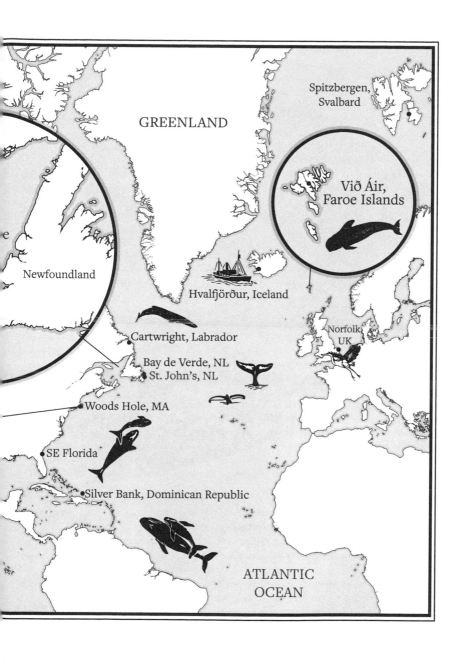

Spitzbergen, Svalbard

GREENLAND

Við Áir,
Faroe Islands

Newfoundland

Hvalfjörður, Iceland

Cartwright, Labrador

Norfolk
UK

Bay de Verde, NL
St. John's, NL

Woods Hole, MA

SE Florida

Silver Bank, Dominican Republic

ATLANTIC
OCEAN

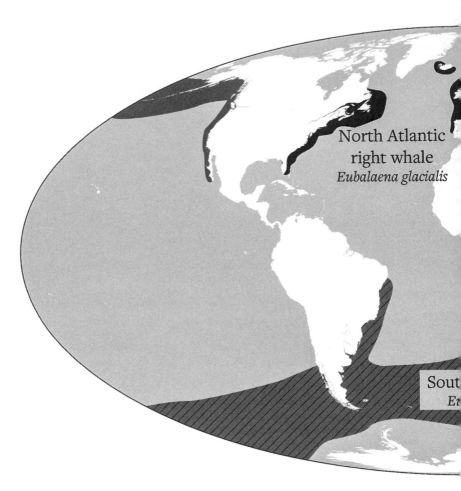

Global distribution of the three species of right whales.
Data: https://simple.wikipedia.org/wiki/Right_whale#/media
/File:Eubalaena_range_map.png. © Plot: Woods Hole Oceanographic
Institution, Natalie Renier, WHOI Creative.

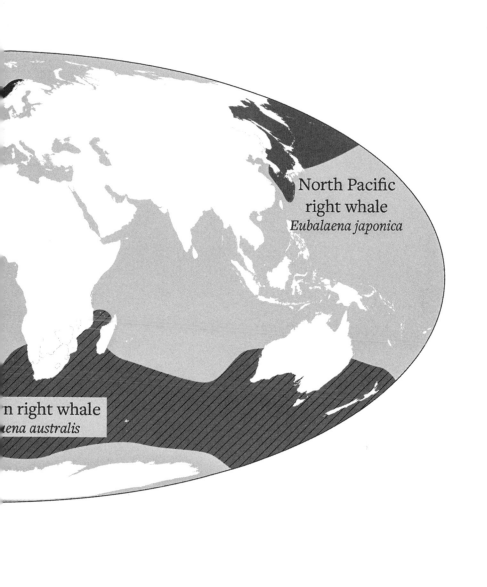

North Pacific
right whale
Eubalaena japonica

n right whale
...ena australis

Preface

"Whaling: The action, practice, or business of catching whales" (*Oxford English Dictionary*).

No consideration of intent or lack thereof.

"Catch: To capture, esp. that which tries to escape; hence, to ensnare, surprise, overtake, reach, get at" (*Oxford English Dictionary*).

Thus, modern industrial practices, such as shipping, trap fisheries, gill nets, and trawls, are all, albeit accidentally, whaling if they catch a whale.

Recently, I spent an early April day in the southwestern corner of Cape Cod Bay, in eastern Massachusetts, in the United States, with a friend. He had been at sea his entire working life, but had never knowingly been close to a right whale. His day job was master of an oil tanker on the Valdez, Alaska, to San Francisco, California, run, where he might have been close to a North Pacific right whale (*Eubalaena japonica*). He was vastly overqualified to skipper our boat, which he did while I piloted a small drone to measure the lengths and widths of the many feeding North Atlantic right whales (*Eubalaena glacialis*) we had found in a small area. There was no wind that day. The sea was like a millpond. It was crisp, cold, sunny, and quiet. We shut down the motor, drifted, watched, and listened. As each animal surfaced, exhaled, and immediately inhaled, we listened to the unique cadence of their breaths, and we watched their steady progress through the water with their mouths wide open, filtering the clouds of food close to the surface. Periodically, they slowly closed in on the boat, and we could see into their open mouths, with small eddies of water peeling away from their lips. Much larger eddies formed in their wakes as their powerful tails and bodies pushed them along. They made tight turns, using their huge flippers and tails as rudders, to keep themselves within the food patches. This went on all day. As the sun started to sink behind the cliffs on the nearby western shore of Cape Cod Bay, their creamy white upper jaws, just visible above the surface, turned to a vibrant golden hue. It was a peaceful, majestic, timeless sight, and a huge privilege to be permitted to study these animals. At the end of the day, my friend said that he understood why I care so passionately for them. Words often fail when I try to express the awe and wonder that these animals elicit; this book is my attempt to do them justice, and keep them out of jeopardy.

My hope is to convince you that the welfare of individual North Atlantic right whales, and the very survival of the species, is in our hands. Few humans eat whale meat anymore, but fishing techniques unintentionally harm and kill whales. Even vegetarians contribute to the problem, as we all benefit from global shipping of consumer goods and fuel, which, in its current iteration, leads to fatal collisions with whales. Entanglement in fishing gear can sentence these animals to months of pain and a slow death. Both the US and Canadian governments are stuck in a major conflict of interest: protecting the livelihoods and businesses of the marine transportation and fishing industries, but at the same time recognizing the value of biodiversity, animal welfare, and avoidance of species extinction. Recently, the latter values have taken a back seat. It doesn't have to be this way. We have the technology and the collaborations that are necessary to change the right whales' future, but consumers have to use their wallets to make it happen. Hopefully, politicians still listen to their electorate.

Though I will use my personal experiences to make this argument, this book is not a memoir. I use descriptions of my life and work, and that of many, many others, to explain basic principles in marine science and what it would mean to lose this and other species. I also explain how we all can help whales to prosper. This story is, at times, gruesome, but I entreat you to stick with it. Again, I believe we can make it right. The fundamental problem for North Atlantic right whales, as for so many of us, is that they can't make an adequate living and they struggle to raise a family successfully. Their carefully evolved energy budget does not work anymore.

Right whales' habit of swimming for many hours at a time with their mouths open to filter food leaves them susceptible to strikes by vessels and to being entangled in rope wrapped around their heads, flippers, and tails. (I use the generic word "rope" here, but once it is assigned a known purpose, rope is more correctly called "line," so in this book both terms will be used as appropriate.) Whales can be found feeding from the surface to the bottom—wherever the food is. Researchers have spotted them with mud on their heads, a sign that they sometimes come into contact with the ocean bottom. Rope entanglement is one of the leading causes of lethal and sublethal trauma in the North Atlantic right whale.[1] Vertical lines used to mark and retrieve lobster and crab traps are the commonest types of rope in the water column of the Northwest Atlantic Ocean, both in the United States (overwhelmingly lobster) and Canada (lobster and snow crab, primarily).

In addition, vessel collisions commonly kill whales. Lethal entanglements and vessel collisions kill them directly—an expense. Poor feeding opportunities (reduced income) and sublethal entanglements and collisions (increased, unbudgeted expenses) reduce the number of calves produced (through adult female ill health)—a net loss. Less income, more expense, and less capital stock leads to a shrinking budget for the coming years.

Like most large whale species, right whales lack teeth. Instead, they have horny plates of a material called baleen suspended from their upper jaws. Some baleen whale species gulp larger prey, while right whales skim their small prey by swimming slowly and steadily. Baleen plates have hairy fringes that make a fine filter, so that right whales can swim through the water with open mouths, sieving through clouds of drifting animals, called zooplankton, that are smaller than rice grains. The water flows out through the baleen, creating endless eddies, while the food is concentrated and swallowed. This is an incredibly efficient way for a very large animal to eat very small ones. These zooplankton, primarily copepods, are oil rich and provide energy for the whales to exist, move, grow, and reproduce. The blubber coats of healthy right whales are full of oil and make the animals buoyant. Rich in oil, slow swimmers, mouths full of valuable baleen, and usually buoyant once they die: these traits made them very early targets for whalers.

Right whales find food mainly in New England and eastern Canada. Their important feeding areas include the continental shelf from New York Bight, across to south and east of Nantucket, and on to Georges Bank, up through the Gulf of Maine to south of Nova Scotia, and into the Gulf of St. Lawrence. Occasionally, they also migrate to forage in the eastern part of the North Atlantic, in places where, centuries ago, they used to be common: the Bay of Biscay, Ireland, and from Scotland to Norway. When, and if, a female has built up adequate energy reserves, she will be fat enough to get pregnant. Some never do. She will remain pregnant for about a year, calving in US waters east of Georgia and Florida between December and March. Centuries ago, females would also calve off sub-Saharan Africa. A colleague of mine, Katie Jackson, described calf behavior:

Once the single calf is born, it suckles regularly, staying close to its mother. Mostly the mother is resting and the calf spends time swimming around, alongside, and crisscrossing under the mother, but especially her head and chin. Body contact is also regularly observed. I'm not sure if this is mother-calf bonding or simply surface resting with intermittent nursing, but you can picture a relatively inactive mother who sometimes changes her orientation (head tilt posture, vertical in water, on her side, or belly up), with a calf maneuvering around her in constant motion. As the calf grows, we start to see more active behaviors like rolling, waving a flipper or tail in the air, and launching itself out of the water.

The calves grow fast, a centimeter or more a day, as they begin the long migration north, while the mother's girth shrinks as she pumps energy into the calf. We can see mother and calf pairs in Cape Cod Bay by April. They continue to suckle for a year or more. In their first year, calves grow from about 13 to 33 feet (4 to 10 m), and then grow more slowly, reaching 46 feet (14 m) or more as mature adults. Females first calve at 9 to 10 years of age, if they manage to get pregnant. They used to be able to produce a calf every 3 years, but lately, changing food availability and sublethal trauma have stretched the inter-calving interval to much longer, even up to 12 years.

Scientists have gleaned this level of detail, from what is essentially a rather cryptic animal, primarily by collecting thousands of photographs of these individually recognizable whales. The photographs are shared with a central database maintained at the New England Aquarium in Boston, Massachusetts, and matched to a catalog of the individuals founded by Scott Kraus and colleagues.[2] Each whale is given a four-digit number. Some are also given a name, usually related to an identifying feature, but occasionally for other reasons. But centuries before we knew them as individuals, thousands were killed for their oil-rich blubber and hugely valuable baleen.[3]

By the twelfth century, Basque whalers from southwestern France and northeastern Spain had grown proficient at harvesting North Atlantic right whale baleen and blubber. This activity was so important to the economy that it led to an early corporate tax: in 1150, King Sancho VI "the Wise" of Navarre allowed the city of San Sebastian to tax baleen.[4] The centuries of whaling that followed led to the depletion of the species to such a degree that whalers no longer saw right whales as profitable targets—by the seventeenth century, they were just too hard to find. Over the next three hundred years, whalers still hunted them when they appeared; the last recorded harpooning of a North Atlantic right whale mother and calf was in 1967 off the Portuguese island of Madeira.[5]

By the early 1950s, researchers thought right whales were close to extinct. Then, in 1955, oceanographer Bill Schevill at Cape Cod's Woods Hole Oceanographic Institution, where I have worked since 1986, described sightings from boats and airplanes of live right whales and made sound recordings. Acoustic engineer Bill Watkins joined him in 1958. They sus-

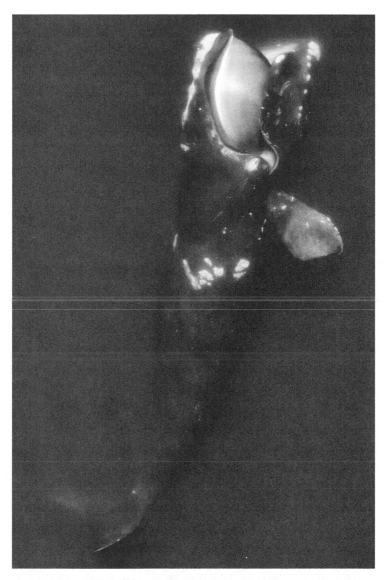

North Atlantic right whale #3760 subsurface feeding on its side in Cape Cod Bay, April 7, 2019. The top of the image shows callosities (patches of thickened, horny skin) on the upper jaw (*left*) and on the lower lip and jaw (*right*). Baleen plates are suspended from the upper jaw. This whale is unique in also having callosity material evident on its right side above the flipper. Callosities are usually limited to the head: jaws, lip margins, and around the two blowholes and eyes. Aerial drone photo on a very calm day: Jacob Barbaro and Hollis Europe NOAA SWFSC/Author. Permit: NMFS #21371.

pected what other researchers have now confirmed, that they could visually identify individual animals by distinct patterns of horny, roughened skin on the head, called callosities. This discovery ultimately led to today's catalog. Scott Kraus, Howard Winn, and many others began to spot and track these whales in the 1970s and 1980s in the waters off eastern Canada, New England, and Florida and Georgia. Essentially, the species had been rediscovered, and by the mid-1980s, researchers had a decent understanding of seasonal right whale movements.

The species then slowly grew until about 2010. The severity of entanglements has steadily increased, however, as a result of modern technology providing stronger and hence more lethal rope, as well as other factors, such as increasing use by right whales of the Gulf of St. Lawrence in Canada, where snow crab traps, with their thicker retrieval line, are set. Vessel strikes have also increased in that area. Thus, the recent growth in the North Atlantic right whale population has reversed. In US and Canadian waters, during the 2017–2020 period—just four short years—10 percent of the species has died. In November 2020, the best estimate for the total number of North Atlantic right whales remaining in the species was a mere 356 animals.[6]

Over the past millennium, the North Atlantic right whale has only just avoided extinction, first from deliberate hunting, and now inadvertently from fish harvesters and mariners. We have reached a tipping point: the right whale species cannot continue to withstand the mortality and morbidity it suffers from fishing-gear entanglement and vessel collision. We must also consider what each individual is going through as it struggles with persistent rope entanglement.

To solve the problem, we need to have the understanding,

commitment, and optimism to carry through with what has to be done—by fundamentally changing fishing and shipping practices. But we also need to make these changes in ways that are sensitive to the lives of the humans that work in vessels at sea and harvest seafood. Both industries have already borne substantial costs in the name of right whale conservation, with inadequate results. Right whales are a special example of mammals that have evolved to thrive in an unforgiving environment and are specialized in diverse and remarkable ways to exploit specific aquatic resources and environments. We must be the same. The challenge is to find solutions that are sustainable, both for whales and for humans dependent on these marine industries.

This is a story that began for me as a child in England, raised by caring people, learning from our challenges and traumas, as all families do. I was taught how to survive on the water, maintain boats, and explore. I trained as a veterinarian, but I also had the chance to pursue my own curiosity. I was shown the enormous wealth of a productive marine ecosystem, off eastern Newfoundland, but also the harsh reality of the trauma whales face when in conflict with humans harvesting a mutual food resource. An opportunity arose to document the remarkable efficacy of direct harvest of large whales in Iceland—a reality whose relevance to my later work took decades to come in to focus.

I then describe a small window I was given into the millennia of native subsistence harvest of the bowhead whale in the Arctic. The native hunters had truly conserved the whales' habitat, and hence the whales, in spite of the best efforts of both nineteenth-century commercial whalers from New England to wipe out the species and recent oil exploration to degrade its habitat. The Alaskan Iñupiaq sense of the long view, and respect for the whales as a part of their culture, made me hope that modern marine industries could also sustainably coexist with right whales in their habitat.

As I slowly grew to understand the impacts of industrial fishing practices and vessel collisions on large whales, I fell into the role of a large whale trauma diagnostician. Along with a few dear colleagues, I provided a scorecard for government efforts aimed at reducing such impacts. We worked with all of the affected whale species in New England: humpback, blue, fin, sperm, minke, and right, in addition to smaller whales, dolphins, and seals. But it was the right whales that were the most prominent and imperiled in their plight. But what is good for right whales will be good for the others, too. We tried to intervene with some sick whales, to reduce their suffering, but realized that prevention of the trauma was the only lasting solution. So now I work with scientists, engineers, fishermen, lawyers, government managers, and nongovernmental organizations to promote a safe, profitable, sustainable seafood harvest that will allow the North Atlantic right whale to turn another corner and prosper once more. Despite all I've seen, I have hope. I believe we can reverse the trend such that a thousand years hence, right whales will be as numerous as before we started killing them, whether with intent or by accident.

1

Young Man, There Are No Whales Left

Whenever I enter an auditorium or arena, I look at the scene and try to decide where to sit. In 2005, I was in the vast open-air arena surrounding the display pool of SeaWorld, in San Diego, California. I had misgivings about being there, given my concerns about how well killer whales were able to adapt to captivity, given their high intelligence, extreme mobility, and strong, cohesive societies. About a thousand of my marine mammal biologist colleagues were gathering. But as my eyes fell on the whales in the pool, and on the growing knots of people in the tiered seating, I retreated to the heights of the back row and found a seat where I did not have to interact with others. I sat and ruminated on the talk I was going to give the next day. I wondered if my data would gain attention. How could I translate the science into meaningful change in the suffering I knew that whales in the wild were experiencing? At least no fishing gear entangled the killer whales on display. Constricting rope was not slowly amputating their flippers. The entangled North Atlantic right whales I was going to talk about the next day were trussed up and in constant pain. Some were dying slow deaths over months. I watched the whales and trainers below preparing for a show designed to reassure the scientists in the audience of the value of the aquarium environment. From where I sat, the trainers were small dots. I could not focus on them.

The title of my talk was "Morbidity and Mortality of Chronically Entangled North Atlantic Right Whales: A Major Welfare Issue."[1] It concerned the enduring conflict between whale welfare and sectors of the growing commercial fishing industry whose age-old techniques were being replaced with stronger ropes that were causing increasing numbers of long-term whale entanglements. I will explain how this all goes down in more detail later, but briefly, right whales swim into the lines used to mark and retrieve lobster and crab traps on the ocean floor. Hitting a line makes them panic, and they tend to spin, trying to evade it, but instead they get multiple wraps of line around various body parts, such as the upper jaw, flippers, and tail. If they cannot disentangle themselves, the wraps constrict through time and slowly squeeze the life out of the animal over a matter of months. I struggled then, and I struggle now, to understand, communicate, and attempt to rectify what I see as a huge injustice to these animals.

My talk was well received, but I knew that I was largely preaching to the choir. I needed to reach out to the trap fishers, their catch buyers, wholesalers, retailers, and consumers. Since then, I have continued to struggle with my knowledge of this problem, my role as a research scientist, and my desire as a veterinarian to also be an effective advocate for individual animals. While I was wearing the dispassionate cloak of academic formality and apparent objectivity—being what I was paid to be in my job—my heart was breaking for what these animals were going through.

I realized that the trauma that I was describing was unseen by most people in the audience, and around the world. They needed to understand the story of the animals, how they suffered, and what we should do so they would suffer at our hands no more.

Unseen trauma and suffering were things I had come to know as a child. I was the youngest of four. My parents were English teenagers in the 1930s, and in the 1940s, my mother was a maternity nurse in central London during the World War II bombing Blitz, my father training to be a medical doctor. By the time I arrived on the scene in 1956, my father was practicing medicine in rural Hampshire, on the English coast southwest of London. As a child, I watched them both care for my father's patients in the village we lived in and those nearby. My father's routine was to do office visits morning and evening at the Surgery, as it was called, and in between he would do his "rounds" in his car, visiting patients in their homes. My mother would help manage the business—the Practice, as it was called. The Practice operated the Surgery out of his partner's house. The actual surgery they did was minor. The Practice had a single phone, and someone had to sit by it, waiting to hear from patients in need and arranging their visits to the Surgery and my father's visits to their homes. Call transfer, voicemail, mobile phones, texts, patient portals, and online scheduling were all unknown at that time. It was simply members of the community with a health care need being treated by the local doctors. Thus, the idea of caring was central to our household. As a teenager, I sometimes did "phone duty," waiting for calls from patients, figuring out where my father might be on his visiting rounds, looking up in the phone book the house he was likely to be at, and calling that house to pass a message on to him.

At times, my mother would talk about her life as a nurse in the Blitz. I remember her describing Hitler's flying bombs that whistled as they approached. It was the ones that ceased to whistle that you worried about, as that meant they were about to land. She told of hearing those bombs and helping a mother in labor to get under the bed, looking for even a token of shelter; of her own mother, in her family home in St. Albans during the war, where there was a daily soup kitchen for refugees. One of them, Edna Bertha Ruppin, or as we knew her, Ruppie, a Jewish woman who had escaped Germany with her husband before the war started, became a de facto member of their family, someone who my mother nurtured till she died.

Both of my parents had been raised in relative comfort. Part of my father's family had come from Norfolk, north of London, and had owned a sailing cruiser there, which they had used for family vacations in the Norfolk Broads in the 1930s. The Norfolk Broads is a series of marshes and freshwater rivers interconnecting expanses of water called broads, confluences of pits created by medieval diggers of peat for fuel. The family photographs show long skirts, jackets and ties, tobacco pipes, and a certain rustic Edwardian elegance joining the vacation scene that had developed in that area. Carrying on in the same, albeit rather less elegant, vein, we had rented various sailing cruisers for vacations in the early 1960s. In 1967 my mother had bought an old Norfolk river cruiser at auction, and that was our getaway for vacations. It was all very English: lots of picnics, walks, dogs, sailing, and endless cups of tea.

However, on vacation in 1968, when I was about to turn twelve, I learned something that changed my life. We kept the boat at a boatyard at an inlet called Barton Staithe on the River Ant, just upstream of Barton Broad. It was where I had learned to sail, in the long-gone wake of a far more illustrious sailing student of the late 1700s, one Horatio Nelson, who would become Admiral Lord Nelson, British naval hero,

who both defeated Napoleon and died at the Battle of Tra-
falgar. Our sleeping habit was to evade the dusk mosquitoes
by anchoring out in the middle of nearby Barton Broad, away
from the marshes, hopefully with a gentle wind. A couple of
days after we had arrived and were sleeping on the boat, my
mother woke me up early in the morning and told me that we
had to go home. I was shocked and disappointed, as those
times on the Norfolk Broads were magical.

My mother explained that my father was not well. Indeed,
he had not been well for a very long time—not since before
she had known him. She told me that he had tried to drown
himself in the broad that night, and that it was not safe for
him to be on the boat feeling like that. We had to move the
boat back to the Staithe, pack up, and drive him to the hospital in London, where he would stay for a while so he could
get better. She told me enough that I realized for the first
time that he was often not very happy at all. I later came to
understand that he was a manic depressive—that is, that he
had what we call bipolar disorder today. But at that point, I
simply understood enough to realize that he had been sick
on and off for my entire life. My overwhelming emotion was
one of deep sadness that I had been so insensitive as to have
failed to understand that my father was mentally unwell, and
that I had failed to help take better care of him. Much later,
as a parent myself, this experience led me to recognize that
young children can blame themselves and shoulder a heavy
burden of guilt—a trauma from an adverse childhood experience they don't fully understand—gaining perspective only in
distant hindsight.

As I carried that crushing weight on my shoulders, we got
the boat back to its mooring and loaded the car. I was dis-
patched to some friends to tell them that we were headed
home and could not meet up with them as planned. To this

day, I remember taking that walk of shame, blinded with tears, to deliver the message, all the while processing what I had just learned from my mother. How could I have been so stupid? How could I have missed it all? As I watched my father with new eyes, the whole pattern became plain. He was gone in the hospital for three weeks. Then he came home, still deeply unwell. Seeing him see me, knowing that now I "knew," my heart was overwhelmed with love and pain for him. Today, I would run over and give him a hug. Back then, in that culture, it would never have crossed my mind.

I now know that he had been depressed as a teenager in secondary school. His old-fashioned English boarding school had been worried enough about him that it had moved him into the headmaster's house for his safety. My parents were married when they were both twenty-five years old, in St. Albans Abbey. My mother's family home abutted the abbey fields, next to the deanery. Many years later, soon before my mother died in 1977, when I was twenty-one, she told me what had happened when her sister was helping her into her dress upstairs, getting ready for the wedding. Her sister was aware of my father's depressive nature and had begged my mother not to go through with the wedding, as she would spend the rest of her life taking care of that sick man. My mother's response was to tell her sister that she understood, she loved him dearly, and that taking care of him was exactly what she was going to do. And she did. Thinking all that through over the years has made me realize that I would not be here today were it not for that very particular love my mother had for my father, just as he was. Her story has helped me as a parent to let my children make their own decisions.

Thus, my childhood was one of slowly learning about my parents, my mother's unconditional love, my father's frailty, and how he turned that disability into strength. In later years, he and I talked some about his disease. That bout of depression in 1968 was in fact a turning point. The drug lithium had recently become available in the UK, and he benefited hugely from it. He never got over his perception of lithium's cost in terms of dulling his mental acuity, but it gave him a relatively stable way forward for the rest of his life. It also emancipated him from electroconvulsive therapy, which had been his only recourse for the previous thirty years. Perhaps the following comment he made about his disease, in the context of telling me that many of his medical patients had as much or more need for mental as for physical healing, was the most striking. He said that his patients asked him why it was that he was the first doctor they had spoken with that truly understood their mental anguish. He would smile, share his past, and keep listening.

My parents championed their children having whatever independence they sought that was commensurate with a modicum of safety. Thus, I found myself at the age of fifteen on the train from Hampshire to Norfolk, to be placed, in theory, under the watchful eye of my eldest sister Liz, who was ten years older than me and living in Norfolk, quite close to Barton Staithe. Thus, she became my second mother. We had a small open sailboat with a boom tent. I would camp and cruise the Norfolk rivers for weeks during my school vacations. I would periodically visit Liz for laundry, dinner, and a bath (showers being nonexistent in England at that time). Having children herself, she had dinner on the table quite regularly, so during one meal, she asked me what I was going to do with my life, a question I really had not considered at that point. But she pointed out that I would need to have an answer quite soon, as applications for university were pending in a year or so. I naively dismissed the idea of medicine, given that my father and brother were both medical doctors and I wanted to be different. Liz then suggested being a veterinarian. Without giving it much thought at all, I agreed. In hindsight, being born into such a nurturing environment was a level of privilege that I did not begin to appreciate at that time. Today I do.

A few months later, I sat down with the teacher at my high school who managed university applications. He asked the big question, and I said I wanted to become a veterinarian. In England, one made such commitments going into the undergraduate stage. The teacher did not question me either, so that was it. I applied to all the vet schools in the United Kingdom, got into some, and accepted an offer from the University of Cambridge.

I arrived at Cambridge in the autumn of 1975. One might think I had a sense of self-worth by then, having been accepted into that prestigious, long-lived jewel of tertiary education, founded in 1209. I was admitted to its second college, Pembroke, founded in 1347. However, any whiff of success was doused within my first days. Walking out of the college front gate at the Porter's Lodge, I ran into a "friend" who had left my high school the previous year and was now a Cambridge undergraduate. The conversation went something like this: "Michael, nice to see you, are you visiting?" "No, I just started at Pembroke." "Really—you were supposed to be stupid." My ego was returned to its habitual place, firmly in the impostor syndrome camp. A few days later, I met my supervisor. (Cambridge has a remarkable tutorial system, whereby each student has a close relationship with a supervisor, a faculty member who tries to ensure that the student thrives.) His opening, admittedly lighthearted, gambit was to tell me that the year that I applied to the university, there had been a memo from the veterinary school exhorting colleges to admit veterinary students, as very small intakes in previous years had threatened government closure of the school. The implication was that without that need, I would not have been admitted. The impostor syndrome devil was growing again.

My first start at Cambridge in 1975 was short-lived, due to a debilitating, persistent bout of mononucleosis. I started again a year later, although I did not shake the disease entirely until the summer of 1978. The silver lining of that cloud was that I spent a lot of time recuperating at home, being cared for by my mother. After a decade of boarding school, it was great to get to know her as I was becoming an adult, learning some of the stories I've told above. Before I began again at Cambridge in 1976, she developed cancer and died the following year. It took me many years to really process her loss. But that time with her was rich with caring. As she died, she was sensitive to the end as to how other people were processing her demise. Worrying far more about others than herself. Asking for perspectives on my siblings, my father, and her many friends and other relations. Able to think and talk quite independently of the intense chronic pain she suffered for nine months. Morphine was an essential part of her routine. She took huge pleasure in the knowledge that her botanist brother Stephen had founded a major medicinal morphine industry in Tasmania, Australia, in the 1960s. Watching what appropriate analgesia can do to manage suffering that was very personal was an important lesson for me. It was another piece of the jigsaw puzzle I was unknowingly building that would lead me to where this book is going.

Back at Cambridge, I found preclinical veterinary medical coursework to be utterly engrossing. I still had no real understanding of why I had elected to pursue a veterinary career, but I found the components of the knowledge required of that path to be deeply interesting. Veterinary science was a unique way to gain a thorough understanding of how biology worked, in the context of disease. I learned much about comparative anatomy, physiology, biochemistry, pathology, and other topics. Much of it was learned alongside medical students, although the anatomy, obviously, was distinct, with more legs involved. But a common thread that bridged these different subjects slowly became apparent to me. The professors and lecturers obviously needed tricks to recapture the wandering attention of their large lecture classes, and a common one was to point out exceptions to the various rules of biology that they were trying to instill in us. The thread that I noticed was that as often as not, these exceptions involved marine mammals. The physiologists talked about breath-hold diving by sperm whales, which could dive for an hour on a single breath. The anatomists described the adaptations in the skin of dolphins to minimize water drag while swimming. The bone pathologists explained that manatees and dugongs had very dense bones to balance the buoyancy of their lungs. I suspect that my failure to really fit in during my earlier schooling made me want to be different, to look for the outlier. Marine mammals were beginning to look like the outliers that I could identify with and wonder about.

Two friends—one a medical student, the other another veterinary student—and I joined the university scuba club. It met weekly for a sandwich lunch. There, the three of us talked about summer plans, and the idea of traveling to the Faroe Islands came up. The islanders had a history of driving pilot whales ashore for food. Our logic was that we could look at the whales and learn from the islanders.

We needed funding to get to the Faroes, which led me to the office of Professor Sir Richard Harrison, chair of the Anatomy Department. When I explained our mission, he responded, "Young man, there are no whales left in the North Atlantic; they have killed them all." Retreating, I consulted with a friend of my father who had served in the British Army, defending the Faroes against a threat of German invasion in World War II. That led to a phone call with the owner of a Faroese fish plant, who laughed at Professor Harrison's original response. "Not sure that is true," the fish plant owner said. "We killed another 350 pilot whales in a drive hunt just yesterday." So I returned to Professor Harrison with the news, and he graciously pulled out the relevant checkbook. My first lessons in fundraising were the value of persistence, and that of the telephone.

Next, I went to various libraries in Cambridge to learn about pilot whales, whaling, and the Faroe Islands. To write proposals to raise money to study the pilot whales, I needed to know what we didn't know. We wondered what the whaling was doing to the species' status: Were there independent stocks of pilot whales across the North Atlantic? It was the first time that I had a truly personal quest—a need to learn because I wanted to find out something not known, as opposed to being told what I was supposed to know that someone else had already figured out. I used the then-novel technology of doing a literature database search, involving a telex to the US and a printed copy of the desired result mailed to me from there. This experience was liberating. I found myself at the "whale" bookshelf at the Scott Polar Research Institute library. The librarian, a friend of a friend, showed me an entry in the library's visitor's book that had been made by biologist Hal Whitehead and suggested that I track him down. Hal was starting a PhD at the Cambridge Zoology Department, studying the biology of humpback whales off the east coast of Newfoundland. He asked me to work for him, but I was committed to going to the Faroes that summer. But I signed on to work with him the following summer, in 1979.

With this ramble through my youth, I hope I have set out the edges of the jigsaw puzzle—the borders, the framework of how I would spend the rest of my life. I had loving parents who were kind to those around them, whether related or not. I had watched my mother die slowly. I had witnessed the huge value of analgesia for mitigating chronic pain, and I understood a little bit about chronic trauma. During the summer of 1979 in Newfoundland, and the following winter in the Caribbean, Hal opened my eyes to the majesty of large, powerful whales and sparked my curiosity about how they feed, breed, and interact with one another and their environment. It was the beginning of wonder and worry about whales for me.

2

The First Whale I Had Ever Seen

Cod traps, first invented in Labrador in the 1860s and widely used in Newfoundland until a fishing moratorium in 1992, consist of a cube or "room" of netting, which fish enter through a door in the front wall.[1] The fish are led to the door by a wall of netting, called a leader, that extends from the shore to the trap, with floats and weights attached to keep it vertical. The trap has corner anchors. The cod trap fishery provided an excellent system for harvesting large numbers of high-quality fish. The traps were set on ledges where capelin and their predator, the codfish, were known to congregate. Given that humpback whales also feed on capelin, there was an inevitable conflict, as the whales and humans indirectly compete for workspace.

On a cold, windy, foggy day in early July 1979, five of us on a small, leaky, but plucky wooden sailboat with heavily reefed sails were making our way into a cove, called the Backside, behind the small fishing community of Bay de Verde, Newfoundland. Each of the cove's many underwater ledges had a cod trap, marked by surface buoys. There were whales in the cove as well. I had been slowly getting to know my crewmates: Hal, Kathy Ingham Pennoyer, Pat Harcourt, and

Hannah Clark. But that day made me realize that Hannah had nerves of steel and was a superb boat handler. There was not much room between the traps, whales, and rocks, but despite the wind and seas, she carefully dodged and weaved between them, and stood the boat up in the gusts as if she had been doing so all her life, which indeed she had. She had an amazing eye for reading the water's surface for clues about what was below. Water in the whales' breaths was swept away as if they were steam engines on tracks. We watched in dismay as a humpback whale got entangled in the anchor line of a trap. After about twenty minutes of strenuous activity, it was able to free itself. There was another whale nearby, and we speculated as to whether it was trying to help the entangled whale get free. This was the first time I had seen an entangled whale. I had no idea then how many more I would see in the future, both alive and dead. The entanglement that we witnessed was a common event at the time we were in the province. As the cod fishery declined and the capelin moved offshore in later years, the inshore entanglement problem diminished, and offshore entanglements proliferated, but at the time we were there, entanglement was a very significant inshore issue. We all felt helpless and left the cove with a sense of foreboding. Conversations about this animal, and other entanglements, kept on surfacing that summer and in the following years. We knew that the relationships of whales and other predators have evolved over millions of years in lockstep with their prey. But when humans enter the mix, seeking to harvest food industrially for commercial profit over and above their communal needs, the system can spiral into disarray, with many unintended consequences.

How did all of us get to Newfoundland, home of fishing families, codfish, and whales? I had spent the winter of 1978 in the

Cambridge Zoology Department, for an elective final undergraduate year, during which I got to know Hal a bit. He had
an office in the basement of the department's museum. We
would sometimes meet at the customary departmental tea
breaks midmorning and midafternoon. Two years earlier,
he had been a scientist aboard RV *Regina Maris*, a 144-foot
(44 m) barquentine, when she was used as a platform for
a group studying North Atlantic humpback whales. They
studied the whales on one of their breeding grounds—a series
of three coral banks north of the Dominican Republic (Mou-
choir, Silver, and Navidad) that is an underwater extension of
the Bahamas—and on their feeding grounds in Newfoundland and beyond.[2] *Regina Maris* was operated by the Ocean
Research and Education Society, a not-for-profit organization based in Boston, Massachusetts. The society employed
scientists to teach marine mammalogy to undergraduate students on six-week-long cruises. While on *Regina Maris*, Hal
and two colleagues (Kathy and Pat) had resolved to acquire
a smaller, quieter, more mobile boat to work from in Newfoundland. They had found the 28-foot (9 m) sailing yawl
Firenze in Halifax, Nova Scotia, in May 1977 and sailed her
to St. John's, Newfoundland. They arrived at about the same
time the whales they had seen on the coral banks were com-
pleting their northward migration.

Hal described to me their summer of 1978 in Newfoundland on *Firenze*. They were based in the harbor of Bay de
Verde, which is some 30 nautical miles north of St. John's,
the capital of the province of Newfoundland and Labrador,
Canada. Their goal was to figure out how humpbacks fit into
the local ecosystem, why they went where they did, what
they got up to when they got there, and how they interacted
with one another, their prey, and humans. Humpback whales
have unique pigmentation patterns on the underside of the

tail flukes, which allow individuals to be repeatedly identified.[3] If you quietly follow them in a boat, you will see that
after they have surfaced to take a few breaths and are ready
to dive again, they arch their backs, and their heads go down.
They then straighten their backs, pivot around their extended
flippers, and push their flukes up into the air in a graceful
curve as they dive, giving the observer a view of those identifying patterns. Thus, we were able to record which whales
were sighted where and when, which opened the door to
individual-based analysis of how many animals were present,
how they migrated, who they associated with, and what significant events were associated with a particular animal.
This technique was being used by an increasing number of
scientists across the region, and it was beginning to show
that these whales gathered in the Caribbean to breed and
calve, especially on Silver Bank, and then migrated north to
fatten up in separate sub-stocks in the Gulf of Maine, the Scotian Shelf, Newfoundland and Labrador, and Iceland. Today
this technique has blossomed into global, web-based fluke
matching systems, to which scientists and many members of
the public are contributing.[4]

Hal was also developing protocols for following individual
animals to learn how they interacted with one another and
how their behavior changed with environmental conditions.
This was the beginning of a productive career that spawned
new and, at times, controversial ideas about the complex
inheritance of cultural traits by some species of whales.

Hal then went on to tell me of the single-handed voyage he
took to survey for whales to the north of the island after the
team's work in the Bay de Verde area that previous summer.
On his way back, he nearly lost the boat when he was
knocked down in a gale and lost both masts off Cape Freels.
He limped into Wesleyville, and the islanders took care of

him. *Firenze* was not a very seaworthy vessel and required a lot of tender loving care. But Hal was a very resilient sailor. Having circumnavigated the island in an even smaller boat a few years previously, he had noted where the humpback whales were most common and decided to return to do his PhD field research there. Using a relatively small sailboat to quietly observe and record the behavior of whales has been Hal's modus ever since.[5] Because small boats are far more affordable than larger research vessels, they minimize the operating costs for long-term studies of whales.

That winter I spent a lot of time in the departmental library, reading some primary research literature in depth for the first time. I remember one paper about the harvest of large whales north of the island of South Georgia (1,000 miles east of Cape Horn) in the early decades of the twentieth century, at the beginning of the Southern Ocean whaling era. The number of whales killed off the coast of South Georgia alone in a very few years was staggering and sobering.[6] I was awed that the ecosystem could support these thousands of animals, and dismayed that humans could kill most of them in such a short period. Little did I know that I was to experience that feeling of awe in the flesh the next summer.

Hal's project, which had some Canadian government support, found me a plane ticket, a damp bunk on *Firenze*, and food. My biggest contribution was helping to keep *Firenze* functional while I learned the tools of marine mammal fieldwork. I also underwent a crash course on socialization in very close quarters, an experience for which ten years in an elite, post-Empire, British, single-sex boarding-school educational system was extremely poor preparation. But that same background had given me access to this opportunity. Today we worry that students seeking marine mammal research experience get little equity of opportunity, given our society's

Hal Whitehead on *Firenze*. Photo: Author.

serious socioeconomic and racial disparities; those disparities existed in the 1970s, too.

I spent June to August 1979 in Newfoundland. Leaving
the island, as I flew from St. John's to Boston, I wrote the following in my journal:

> At take-off I almost want to cry. Something has happened.
> I do not want to leave. It feels like home and it is beautiful,
> even though it is raining. I love you dear old Newfoundland. I
> feel I should now write something stupendous: I've changed
> this summer. I know myself better—for which I have a lot of
> dear friends to thank. . . . Also—meat is less, far less impor-
> tant now.

What had happened? The final statement was easy to explain.
Hal is a vegetarian, and we had eaten an awful lot of lentils,
brown rice, and cabbage salad with Catalina dressing. But
calling the island home takes a bit more explaining. My expo-

sure to wildlife by that point had been mostly in the Norfolk Broads—ducks, geese, swans, great crested grebes, kingfishers, and freshwater fish. Beautiful, appreciated, and special, but I was completely unprepared for the sheer power and mass of the throbbing sea life I witnessed off Bay de Verde with Hal and his crew on *Firenze*.

As soon as I had finished my exams in Cambridge, I had flown to Boston, but before going on to Newfoundland, I had visited with some colleagues of Hal's, Katy and Roger Payne. They were the whale biologists who had helped introduce the world to humpback whale songs, recorded by a US Navy scientist off the island of Bermuda and released on a long-playing vinyl record in 1970. The record, thanks to the Paynes' publicity efforts, became the best-selling environmental album in history and has been described as "the soundtrack of the 'Save the Whales' campaign."[7]

I told the Paynes about my experiences in the Faroe Islands the previous summer. At that time, the only whaling thought to occur on the Faroe Islands was a drive fishery for pilot whales that had been active since Norse peoples settled the island, as evidenced by a Faroese law in 1238 that pertained to the hunt. My intent was to gather information from the animals, once dead, to better understand the stock structure of the North Atlantic population of the species. Roger questioned me closely about the details of the hunt, in which the whales are herded to shore by boats, then killed. He would not be the last to worry about the animal welfare aspects of this hunt. I emphasized that the killing method appeared quite rapid. This had been my first experience of dying whales.

The day my friends and I arrived in Torshavn, the capital city of the Faroes, we heard of a fin whale that had been harpooned and was being towed ashore. We traveled to the site of the old whaling station at Við Áir, on the island of Streymoy.[8] A local fisheries biologist drove us along narrow cliff roads in the rain while I used a 12-inch ruler to push the wiper blade back and forth, as the wiper motors were dead. The steep, wet, winding roads led me back to conversations we had had before we had left Cambridge, in which we had been assured by the secretariat of the International Whaling Commission that pilot whales were the only whales hunted from the island. But there it was, a dead fin whale, being hauled ashore from a fishing vessel, the *Hvitiklettur*, that had an old harpoon cannon welded to its bows. It was the first large whale I had ever seen.

The whaling station had been dormant since 1968, but a new whaling effort had just begun and would continue until 1986. The station is now a museum. Watching this whale being taken apart by Faroese who had worked at the station a decade before was also the first time I had seen a large whale disassembled. The equipment had not been maintained, and cables broke and winches failed, but the local community was pleased to take whale meat home to their freezers, often in drooling laundry baskets on the roofs of their cars. My mind was all of a jumble. The anatomist in me was fascinated, but the budding veterinarian wondered what it took to kill a whale with an explosive harpoon. Little did I know that in a few short years, I would be at the pointy end of that question. My notes list the following whales killed in the Faroe Islands in 1978: 7 fin whales; 70 killer whales, driven in Klaksvik on June 30; 611 pilot whales on three drives (176 on April 1, 97 on August 8, and 338 on August 28); and a bottlenose whale on August 22.

Listening to my story, I could sense that Roger was looking at me somewhat askance, not quite approving of my descrip-

tions of the details that had struck me, wanting me to condemn the practice of whaling. I had no idea at that time that
I would come to take a far broader view of what constituted
whaling than most people did then—a view that eventually
had a significant effect on my ability to condemn harpoon
whaling, once I began to realize that the critics of such practices were themselves consumers of products that had far
slower whale deaths as a by-product.[9]

Before I left the Paynes' house, we talked about the project
that Katy and Roger had started to study the Southern right
whale (*Eubalaena australis*) in Golfo San José, on Peninsula
Valdés in Argentina. They had the idea that right whales
could be identified as individuals using the details of their
callosity patterns, which seemed to be unique to each individual.[10] If they could identify individuals, they could begin
to understand the population dynamics of the species as
well. Roger offered me board and lodging to work at their
very remote "Whale Camp," on a beach near a huge cliff
that allowed unparalleled opportunities for observing whale
behavior. Roger's plan was that I would quit veterinary school
and work there. He almost successfully derailed my veterinary career, but for some reason still unclear to me, I declined
his offer, and did so again when he repeated it in Cambridge
when we met up in the fall of 1979 with Hal—not on a blue-
sky New England late spring day this time, but with damp,
cold drizzle coming in off the Cambridgeshire fens. I was
sorely tempted, but by then I was deep into cow pathology,
and I stuck it out. It was not until 2009 that I was to eventually spend time at Whale Camp, helping to investigate a
major die-off of Southern right whales. Had I known what I
was turning down, my answer could well have been different:
Southern right whales routinely bring their calves up to the
beach in front of the camp.

Leaving the Paynes, I flew to St. John's via Halifax. There was no airport security to speak of at that time, so while I was waiting at the gate in Halifax, the public address system asked me to go the information desk, where I met a boat rigger who had supplied new masts to Hal to pick up a box of rigging parts. I looked in the box and noticed some essential pieces required to make the masts stand up. Knowing that Hal was already at sea at that moment, it was the first time I wondered at Hal's extreme capacity to survive at sea, but not the last.

The road to Bay de Verde from St. John's starts off along the eastern terminus of the Trans-Canada Highway before turning right and running up the northwestern shore of Conception Bay to the outer end of the Bay de Verde Peninsula. The road finally curves downward into the town and small harbor of Bay de Verde, passing the church and continuing down to the fish plants along the east side of the well-enclosed harbor. *Firenze* had just arrived from Wesleyville, so I set to, installing the new hardware to get the rig set up on the boat. The new boom—the piece that runs along the base of the mainsail—was too long. Hal put a hacksaw in my hand and asked me to shorten it suitably. No messing around. I gradually became all too familiar with the somewhat idiosyncratic single-cylinder diesel motor, well-irrigated bilges, and other dubious characteristics of the good ship *Firenze*. In that first week, we initially saw no humpbacks—just a couple of minke whales, along with cold, drizzle, fog, and icebergs. To give the whales time to arrive, we headed north across neighboring Trinity Bay to the Trinity Inn to visit the innkeeper, a local whale acoustician. Sailing back from there in glorious sunshine, we passed seaward of Baccalieu Island, close under the cliffs. There were clouds of birds wheeling around the nesting ledges: puffins, kittiwakes, gannets, fulmars, guillemots, murres, razorbills, and herring and greater black-backed gulls.[11] Today the island is an ecological reserve and supports the greatest diversity of breeding seabirds in eastern North America. To me on that first day, it was an unbelievably beautiful, energy-filled place of wonder. I have been back there in more recent years, and it is always with a sense of trepidation that I wait to see how the bird cliffs have fared, as the island represents my first exposure to the awesome potential of nature to thrive if given a chance.

Still no whales, so we went around to Old Perlican in Trinity Bay to haul the boat and fix some leaks. We also made a visit to a large warehouse building, the Baccalieu Club, a dance hall with a large bar and many regulars. Meeting the local fishing families at the club, with time to talk, we learned much about their lives, history, and current challenges. By then Hannah had joined us. She tried to teach me how to dance. It was the beginning of a relationship that, after forty-plus years, has yet to end. We got married five years later, and I owe my career to her encouragement.

On the way back from Old Perlican to Bay de Verde, we saw our first humpback whales of the season, including a mother and calf. My journal entry for June 16:

> The season has really started now—capelin [a small bait-fish[12]] spawning on the beaches, gannets, puffins, other alcids and kittiwakes feeding in the tickle and whales.

A tickle is a short, narrow channel, and this one, the Baccalieu Tickle, was the channel between the end of the Bay de Verde Peninsula and Baccalieu Island. I wrote of the most dramatic sighting, right under a cliff, of what seemed to be a rock awash, then suddenly morphed into a small humpback lunging through a school of capelin. We watched it switching back and forth between two schools of capelin, wondering if it was giving each school time to re-form before coming back for more. Up close, we watched it turn to lunge, with a sparkle of capelin jumping before it. At the bottom of the journal page, I wrote "GOOD DAY."

The summer slowly evolved into a routine. At dawn, Hal would walk to the top of the hill in Bay de Verde to see if the fog had cleared from the tickle. If it had, or he thought it would, he would leap onto the deck with a loud thump to stir his crew, drag me out of bed, and have me steer the boat out of the harbor, and we would look for whales. Sometimes we would run a standard transect line around the island to survey whale densities; other times we would follow a single whale or group of whales all day to record their behaviors. All the while, we would be taking fluke photographs for identification. We were often working in windy conditions, and it was always cold and wet. To shorten our morning commute, we also laid a mooring in a small cove (Lunen) at the northwestern end of Baccalieu Island, tucked in under a cliff full of puffin burrows. To this day, the memories of watching those puffins come and go have great clarity. On foggy days, or with a gale blowing, we would organize our data, restock the boat, and perhaps explore the barrens between Bay de Verde and Split Point, at the south end of the tickle. One day Hannah and I went for what, for me at least, was a romantic walk. However, I had been full of the joys of my current girlfriend in England, which rather complicated things.

One other strong memory was the day that Hal decided it was time to gather data on the diet of the humpbacks. It was somewhat obvious that they were primarily eating capelin, but we had to know. He had asked me to be sure to bring a wetsuit, but it wasn't that thick, and there were lots of icebergs. The plan was that I would jump into the water with fins,

The crew of *Firenze* in Lunen Cove, Baccalieu Island, Newfoundland, 1979.
Clockwise from left: Hannah Clark (now Moore), author, Nicola Davies, Hal
Whitehead, and Kathy Ingham Pennoyer. Photo: Patricia Harcourt.

snorkel, and an underwater camera and photograph whales
going through schools of capelin with their mouths open. I
did get a couple of reasonable shots, but after the second pass
of a whale only a few feet away from me, I declined to continue my observations. The whales did not seem to care that I
was there, which was both good and bad.

On working days, we would leave the harbor by 6:00 a.m.
and would often not be back before 10:00 p.m., so we took
turns sleeping during the day. Perhaps my strongest memory
of all is sleeping in the early afternoon and dreaming of
whales blowing all around the boat, only to wake, poke my

head out of the cabin, and realize that was indeed the reality. That moment of confusion between dream and reality is unique.

A few days later, I saw live fin whales for the first time. Their speed and grace was remarkable. Thinking back to the one killed by the *Hvitiklettur* in the Faroes brought fresh focus to what it meant to kill a whale. One by one, they were gone, never to return. The ease with which a species could go extinct was obvious.

My sense of how the summer evolved that year for the fish, whales, and birds was that in early June there were few whales, and the seabirds were mostly in their colonies. The seawater was 36°F–40°F (2°C–4°C), and not much was going on. By late June, the capelin spawning induced a rapid change, with improved codfish harvest, water temperatures of 40°F–50°F (6°C–8°C), many more humpbacks, commonly lunge feeding, along with many species of seabirds also feeding on the capelin. Then, in July, once the capelin had spawned, the whales stopped lunging after them at the surface and switched to herring, diving deeper and less predictably.

Once the season began to wind down, we sailed the boat back to St. John's, where we spent a week writing up the data at the Department of Fisheries and Oceans laboratory. I then made an abortive attempt to work with another sailboat to study pilot whales. When that effort did not work out, it was time to head south and home toward England. Back in St. John's, I visited a colleague of Hal's, Jon Lien, who studied animal behavior at Memorial University and had started researching the whale entanglement problem.[13] His interest in birds had been interrupted in 1978 by a plea from a fisherman who had been trying to free a humpback entangled in his net for the previous three months. At that time, there was no organized response to such an event. After Jon and a colleague freed the animal, and thereby not only saved the whale, but also minimized the damage and cost to the fisherman, word spread around the industry, and Jon became heavily involved in the problem. He set up the Entrapment Assistance Program, whose title encapsulated his goals of assisting both the whale and the fisherman. Ultimately, the idea of disentangling whales grew from this beginning into the global effort it is today. The Global Whale Entanglement Rescue Network of the International Whaling Commission works with leaders of established national programs to establish best practices for whale disentanglement, and provides training where needed to address the growing entanglement problem worldwide.[14] When I spoke with him, Jon reported that there were currently more than five hundred humpbacks in Bonavista Bay, which is two bays north of Bay de Verde, of which six had been entangled on the previous day alone. The best way to conclude this account of our visit is a further journal quote:

> Poor man is going spare, but it is so good that someone has done so much in such a short space of time. They developed entanglement release techniques with tools that fishermen can use. They have started teaching them how to release the animals. They are planning devices to reduce the damage to the nets. Clangers and acoustic pingers to warn the whales of where the traps are, and time release buoys for gill nets or maybe acoustic releases. It's good that he has the resources. He is a great man.

Our crew in Bay de Verde had been testing the clangers that Jon was devising as part of our work that summer. The need to mitigate entanglement continues to this day. Jon's ideas and work were fundamental and visionary.

Jon was ultimately awarded the Order of Canada, and was recently the subject of a touring play, *Between Breaths*, that traveled around Canada.[15] Sadly, Jon died on Wednesday, April 14, 2010. The day he died, I was at a week-long meeting of whale entanglement specialists in Kihei, on Maui, in the Hawaiian Islands. At the beginning of the week, we had all been invited for daily early morning paddles with the Kihei Canoe Club. By Thursday, I was the only visitor still showing up. As the club members were launching the outrigger canoe, I realized something was different: there were baskets of flower petals on board. This was a ceremonial paddle to sprinkle petals in the water to remember one of the crew who had recently passed away. I offered not to go, but they insisted that I come with them. It was a beautiful dawn, and we all sprinkled petals for their friend, and I added a few more for Jon. It seemed right.

The whales, too, were leaving Newfoundland in late August, many of them heading south to the Caribbean, especially Silver Bank. I went home to England to regroup, and agreed with Hal (and the university) to defer veterinary school for a year and to help get a boat to the Caribbean so we could do a winter of research on Silver Bank before another summer in Newfoundland.

We needed a boat that was more suitable for offshore work than *Firenze*, which meant delivering *Elendil*, a 33-foot (10 m) sloop, from Brittany, France, to the Caribbean and a rendez-vous with *Regina Maris* on Silver Bank. After getting *Elendil* to the Canary Islands, I spent a month fixing it up. Then Hal, Kathy, and I made the passage from the Canaries to Antigua, and continued on to Silver Bank. My strongest memories of the crossing were of endless days of sun, intensely starry nights, following wind and seas, and magnificent frigate birds high above, endlessly soaring. Long downwind sailing passages are mostly about avoiding chafe—between different crew members as well as line and sails moving back and forth over hard spots. Both can be avoided with endless attention to detail. Personally, it was time to relax, reflect, and begin to process the loss of my mother, which I had parked in a denial box in the depths of my brain for the intervening two years. Reprocessing all of that took time and emotion. But my overall sense was that in contrast to the previous decade of boarding school, I was now in an environment of my choice with people that I wanted to be with. It made a world of difference.

On January 15, 1980, we first anchored on Silver Bank, behind *Regina Maris*, in a big swell. Needing more shelter, we moved up into the windward edge of the bank, which was a series of tightly spaced coral heads with areas of sand between, to get out of the swell. We were able to find spots of flat water, despite the swell breaking on the windward edge of the reef and its precipitous drop-off only a few hundred feet upwind. We snorkeled in the reef over huge brain corals, under a subsurface canopy of staghorn coral, with numerous parrotfish and other species moving in and out. My journal claimed "a completely breathtaking, exciting experience: quite amazing to be at rest in a peaceful anchorage, 70 miles from land." So here I was, a few short months after leaving Newfoundland, at the northern end of the major tropical part of the North Atlantic humpback whale's range, being given another unique glimpse into the vast beauty of their world.

The wintering grounds are where humpback whales go to calve and mate. The whales were there ahead of us, and when we arrived, we found mothers, calves, and single males singing. But something else struck me immediately, as I noted in my journal:

> Everything is very relaxed down here: no shrieking distressed blows, as we would see in Newfoundland after a hardworking foraging dive. Here they do not seem to be exerting themselves. No dramatic fluke ups, just gently lifting a fluke a bit and sliding under.

In these calving areas, solitary humpback males sing the songs first described by Katy and Roger Payne from Bermuda. We could hear them easily with a hydrophone, but if the boat was close to a singer without the hydrophone, I could feel the vibrations of the song through my bare feet on the cabin floor. A number of times, I would be asleep in the forecabin, dreaming of a singing whale, only to be awakened by it and hear and feel the song as it roused me to consciousness.

On January 24, we came across a newly born humpback calf with wrinkled gray skin and curled-up flukes and dorsal fin. My journal picks up the story:

> Unable to swim unaided, the mother supported it on one of her long flippers. There were three other whales escorting the mother and calf pair. Over the next four hours it gained mobility, independence and speed. The mother changed from close attention to the calf to more demonstrative interactions with the escorts. Once, as she dove, with the calf sliding over her rostrum, she expelled a white foot-square patch of possibly placental membrane, about six feet under. I saw it from the foredeck, so dove in to preserve it for later analysis. Now anchored in the deep part of the bank, not much wind and lots of singers, very beautiful, very tired.

Songs can be heard through the hull. This is a very special place.

As days went by, we saw large groups of animals, seemingly focused on a mother with a calf, moving fast and interacting intensely with one another.

On January 31, we met up with *Regina Maris*, and I traded places with Kathy, who had been on *Regina* since we had arrived on Silver Bank on *Elendil*. A change of scene would be good. Working on *Regina* was very different. The whales seemed to be more disturbed by the much larger vessel when it was motoring, though not so much when it was under sail, and the data we were able to collect were less behavioral and more of a census, including some fluke photographs. Census work was best done from the royal yard, which suspended the topmost of five square sails on the foremast. The view of whales from 95 feet above the water was remarkable. But overall, it was harder to get a "feel" for the whales from the larger vessel. On March 5, after a trip into San Juan, Puerto Rico, to drop off a group of students and pick up a new bunch, we met up with *Elendil*, and I transferred back aboard. My journal speaks of the excitement of being back on the smaller boat:

> The most dramatic thing seems to be that Hal & Co have fig-ured out that the fast-moving groups do seem to involve a focal mother and calf with escorts keeping other potential escorts at bay, by lunging, charging and blowing bubble-streams.

In other words, most of the members of the fast-moving groups seemed to be males competing for the mother's attentions. The end result of Hal's observations of the chasing groups was that he and Peter Tyack, who had simultaneously undertaken similar work on the Pacific humpback calving area around Hawaii, co-authored a paper titled "Male Competition in Large Groups of Wintering Humpback Whales."[16]

We continued to collect data, and we periodically rendezvoused with *Regina* for fresh food and mail. Our mode for exchanging goods and people with the ship was rowing a small inflatable boat across the swells. On these trips, the slicker one's boat handling skills, the less wet one got.

Navigation required a combination of recognizing the pattern of visible coral heads, watching the change in the depth of submerged coral heads on the depth sounder, and sun and star sights with a sextant. There was no GPS or Loran at that place and time. Hal had an extraordinary sense of location, and as time went by, he got really good at getting to where we needed to be. The experience of being on the bank, with the humpbacks calving and chasing around, and the coral, and the skies, was indescribably beautiful. This from my journal of March 30:

> Climbed to the crosstrees. Midday, high sun, wind blowing force 6 [25 knots, 45 km/hr]. Fantastic colors: creamy white breakers, psychedelic cauliflower coral heads and azure sky. Too windy to work.

On April 6, we returned to Puerto Plata in the Dominican Republic to finish the season, then headed off in different directions before meeting up again in Bay de Verde in June. What had I learned since arriving in Newfoundland the previous June? I had seen humpbacks in their foraging habitat around Bay de Verde, feeding intensely on capelin and herring, in the enormously productive ecosystem on the east coast of Newfoundland. I had seen the same whales on Silver Bank, calving, singing, and competing for mates. I had been able to watch the lives of humpback whales through the span of a year. It was an enormous privilege and had fundamentally changed my understanding of the natural world. I had also started working on a paper with Hal titled "Distribution and Movements of West Indian Humpback Whales in Winter."[17]

My second summer in Bay de Verde was much like the first, albeit with more entanglements. A drowned humpback whale calf wrapped up in rope at Grates Cove, which is just north of Bay de Verde in Baccalieu Tickle. An adult humpback wrapped up in a huge amount of gill net, taking a week to die. In a strange way, I knew I had to go back to England and complete my veterinary degree, so I tried to push those events—or cases, in medical parlance—into a dark corner of my mind and let them sit quietly. However, they kept on percolating out of my subconscious and dreams into a conscious argument within my head as to what I could usefully do about them.

3

Whaling with Intent

What is worse for a whale—an explosive harpoon or entanglement? One of the big surprises of my life is that the answer to
this question is neither obvious nor clear-cut.

By the time I entered my last year of veterinary school, I had a
strong sense that I wanted to merge my veterinary and marine
mammal worlds. One day in late October 1982, I played
hooky from veterinary school and worked with a local veterinarian to examine a group of pilot whales that had stranded
in the Wash, a large, very shallow embayment with complex,
drying sandflats about two hours' drive north of Cambridge.[1]
This interaction led to discussions with staff at the International Whaling Commission, which is based near Cambridge.
The commission was founded in 1946 to "provide for the
proper conservation of whale stocks and thus make possible
the orderly development of the whaling industry."[2] The Save
the Whales movement in the 1970s, powered in part by the
humpback song recordings, significantly affected the commission. Conservation and whaling have jostled for space on
the commission's agenda in recent decades, but that agenda
has included a growing focus on measures such as mitigating
vessel strike and entanglement impacts on whales. The com-

mission has also been under long-standing and significant pressure to understand the use of harpoons in the context of animal welfare concerns.

At the 31st meeting of the commission in 1979, there was a resolution to establish a whale research center to enhance scientific research on the whales harvested by Iceland. Support was to be provided by the Icelandic government and the whaling company, Hvalur HF. Over the winter beginning in 1982, I met with commission staff at its office to discuss a project to study explosive harpoons in the Icelandic fin whale fishery. The commission needed a veterinarian who knew something about normal whale behavior. My scant qualifications were that I had seen a dead fin whale in the Faroes, as well as some live ones in Baccalieu Tickle, and that I would (hopefully) become a veterinarian in May. I did have a pretty broad understanding of the range of behaviors one might expect from another baleen whale species, the humpback.

I committed to undertake that project with a collaborator from the United States during July 1983. Thus, my first job as a veterinarian was to observe fin whales being killed. Not exactly what I might have envisaged when I applied to veterinary school at the age of sixteen, but it seemed like a reasonable thing to do at the time. I had developed a visceral reaction whenever I thought about, or especially saw, marine mammals: their taxonomy, anatomy, ecology, life histories, and perhaps most especially, their physiology—in particular, how they were able to breath-hold dive for as long as they did—were fundamentally exciting to me. That excitement and awe had not included a curiosity about how they could be killed, and I struggled with the sense that I was entering unknown territory, but the opportunity was there, and I would be paid to do it. To explain the significance of this experience in the broader context of my later exposure to whales dying very slowly from entanglement in fishing gear, I will need to explain in detail how whalers kill whales intentionally. It's gruesome, but the scale of the engineering involved, its relative efficacy, and the skill required to deploy it are remarkable. That perspective was reasonably straightforward. But the broader topic of whaling was mired in a vast controversy as to whether the whalers should be doing what they did at all, and in all of the arguments that rotated around animal welfare, conservation of biodiversity, whether humans should eat meat or not, whether whales are sentient beings, and many other complex nuances. As a carnivore, could I, should I, distinguish between what I had seen in an abattoir, as a veterinary student being trained to inspect meat for any public health risk, of the efficacy of killing domesticated animals, versus that of hunting much larger wild animals? I tried to go at the whole problem with an open mind and focus on the job I had been asked to do. As I slowly got into the details of the system and the challenge of evaluating it, I found myself putting a hold on all the bigger ethical and moral questions and focusing on the details.

At that time, I had no understanding of the significance of this opportunity. I had no knowledge that I would spend a considerable part of my later life examining and worrying over whales killed by ships and, most especially, by rope entanglement. In hindsight, this opportunity served as the positive control of an experiment that has occupied me for much of the past thirty-five years: a test of how well, versus how badly, humans can kill whales. It grounded my perspective on the range of human-induced whale morbidity and mortality.

Human consumption of whales goes back millennia, first by exploiting whale carcasses that had washed ashore. People then started to spear whales near shore in the hope that they would wash ashore. By AD 1000, whalers were harpooning whales from small boats and towing them ashore. Later, they were voyaging greater distances in larger ships, but still towing the whales ashore. As the centuries went by, the techniques, tools, and skills of whaling evolved. Two major changes were the late nineteenth-century Norwegian invention of the explosive harpoon, and the practice of inflating the carcasses with compressed air, which allowed the faster, denser whale species that sink at death to be hauled back to the ship. By the time I arrived on the scene, the Icelandic whaling operation was using fast steam-powered whale-catching ships with 90 mm harpoon cannons mounted on the bows. The whales they hunted were mostly out in the Denmark Straits, west of Iceland, so the whaling ships would steam out 150–200 nautical miles toward Greenland, kill two whales, and then tow them back to the shore station to be hauled up the ramp and "flensed"—their bodies taken apart. Speed was essential once the first whale was killed so that both could be processed before significant decomposition had occurred.

During the previous winter, I had spent a lot of time looking at the many, many reports written as part of the International Whaling Commission documentation process in the previous decades, looking at how harpoons worked, how they could work better, and what their effects were. My strongest memory of that reading was learning about the huge and futile effort to use electrocution as a whale-killing method. With the inevitable high-voltage cables running from the ship to the harpoon, it is perhaps no surprise that it seemed as though more humans than whales succumbed.

By the time I arrived in Reykjavík, I felt as prepared as I could be, but really had very little idea of what I was in for. Hvalur HF and the Marine Research Institute in Reykjavík

Satellite image of the whaling station in Hvalfjörður, Iceland. Credit: Google Earth © 2020 Google Image © 2020 CNES/Airbus.

had agreed with the commission that the project should be undertaken. Indeed, I had been met at the airport by a member of the family that owned the company. That level of hospitality continued for the entire six weeks of my visit. But I feared that there would be a fundamental tension in our relationship, as I was being asked to form an opinion as to how well the whale gunners did their job.

The whaling station was 15 miles up the otherwise largely undeveloped north shore of Hvalfjörður, the first large fjord north of Reykjavík, and a bit more than an hour's drive from the city. It consisted of a dock for the whaling ships and a small cluster of industrial buildings around a ramp. The whales were hauled from the water and up the ramp, which broadened into a large flat surface used for disassembling them. Workers efficiently peeled the blubber and meat off the animals using hooks on cables, directed via suitably located turning blocks to steam-powered winches, and long-handled, curved flensing knives. The station was also equipped with large reciprocating saws to reduce the skeleton into manageable pieces. All of the materials were sent down chutes to processing areas in floors below the flensing deck.

I met with my American collaborator, Richard Lambertsen, another veterinarian. He had been studying the health of the harvested fin whales, focusing especially on a large nematode worm that affects whales' kidneys and associated blood vessels, which he believed could be a significant source of mortality for juvenile fin whales. He was experienced in examining these animals as they were disassembled. The arrangement was that he would examine each fin whale for evidence of the trauma caused by the explosive harpoon. In this way, he could assess the damage that had been done to the vital organs of the animal and infer from that how long it had taken for the animal to die. For instance, looking at one animal in our study, he saw that the harpoon had passed through the left chamber of the heart, and thus concluded that the whale had died rapidly; in another case, the first harpoon had lodged in the whale's back muscle, and the animal had swum for many minutes before being killed by a second harpoon. My role was to acquire data that would provide a behavioral perspective on the question of time to death. I would accompany the whalers on the *Hvalur 6*, watch the whaling equipment and the behavior of the whales, and estimate how fast the whales died.

The harpoons, which weigh 154 pounds (70 kg), have four folding flukes and a 22-pound (10 kg) cast-iron fragmentation grenade containing 1.4 pounds (0.65 kg) of black powder explosive. The grenade is detonated by either of two fuses, one timed to go off very shortly after the harpoon penetrates

Icelandic whale catcher, Denmark Straits, June 1983. The crew hunt for whales from the flying bridge. When the skipper is ready to shoot, he runs along the catwalk to the bow to operate the harpoon cannon. Photo: Author.

the animal, the other triggered by the act of penetration. The harpoon is designed to go deep into the animal before the grenade explodes, at which point the grenade shrapnel has the greatest chance of inflicting mortal damage. The brain of a large whale is a relatively small target and encased in heavy bone, so the most practical target is the chest, whereby the shrapnel can lacerate vital structures such as the lungs, heart, and great vessels. Thus, the gunner aims for the whale's armpit, which overlies those structures.

Preparing harpoons. Photo: Author.

A nylon line (the forerunner) runs from the harpoon to a steel cable connected to the ship's main winch, mounted on the deck in front of the cabin. The forerunner, and the cable, once the harpoon has been fired, turn around a series of pulleys that route them up the ship's mast, around a pulley on the mast, and then back down and thence to the winch. The pulley on the mast is itself suspended by a cable that runs through another pulley nearer the top of the mast before descending belowdecks, where it is attached to a large accumulator spring. This arrangement gives the whole system the elasticity needed to absorb the shock loadings of a whale reacting to the penetration of the harpoon, as well as the detonation of the grenade. A whaling ship could be likened to a 170-foot (21 m) fishing rod, in that just as a fishing rod bends to avoid the strain of the swimming fish breaking the line, the ship's rigging transfers the strain to the accumulator spring. Furthermore, just as a fisherman plays a fish using the reel to tire it until it can be reeled in, a whaling ship has a winch operator who lets the cable spool out while playing the brake, hauling in once there is less tension on the line as the whale tires. Obviously, if the explosive harpoon were to kill the whale instantaneously every time, this complex shock-absorbing system would not be needed. If a whale was not killed by the first harpoon, but was nonetheless well attached to it, and thence to the ship, an 88-pound (40 kg) "killer" harpoon, also with a 22-pound grenade, but not attached to the ship, was fired after the whale had been hauled close to the ship.

Once the whale was dead, the crew hauled it under the bow, made an incision in the belly to allow the blood to drain and cold water to start cooling the carcass, and then maneuvered the carcass so as to pass a large padded chain around the peduncle, which is where the tail flukes attach to the body. They then passed the chain through two holes in the bulwarks at deck level, with a clamp to hold the tension, and cinched the chain tight with the steam winch. Once the tail was secure, they made a series of cuts in the peduncle, which consists of blood vessels, bone, and tendons. These cuts exposed a large, thick-walled vein that in life drains the blood from the flukes. They incised the vein to insert and tie off a rubber hose, then pumped ice-cold seawater through the hose to cool the carcass and improve meat quality. With the first whale thus tied to the side of the ship, they towed it by the tail while killing the second whale. The ship then towed both whales back to the shore station for processing.

The data I was to focus on included time from when the harpoon was fired to when the whale's mouth slackened, its flippers relaxed, and all movement ceased. Then, once it was ashore, Richard Lambertsen would determine the location of the harpoon and the surrounding damage.

Once I started to appreciate the rhythm of the process, I began to recognize the challenges that I faced. I had a cabin close to the wheelhouse, from which I was able to watch the hunt. The hunt was conducted from the open flying bridge above the wheelhouse, where the skipper, who was also the gunner, would work with his crew to spot and approach a whale. During the hunt, the ship was steered from the flying bridge. At the appropriate moment, the gunner would come rapidly down the ladder from the flying bridge, sprint along the catwalk to the foredeck, unleash the cannon, and aim and fire at the whale when it was in range and a suitable target was exposed. It could take many hours to get into the right position and kill a whale. From the beginning of the hunt for the first whale to the demise of the second could take 18-24 hours. My first challenge was to stay awake at my position in the wheelhouse to be ready to take photographs, start the stopwatch, and record the key time points. Not having anyone to relieve my watch, I sometimes found myself dozing in a swivel chair by the ship's main wheel, stopwatch in one hand, camera around my neck, only to be awakened by the cannon firing—there was no possibility of sleeping through that alarm. In those cases, the stopwatch may have started a few seconds late, but I still got the data. Then I would sleep for 18 hours on the way back to the station.

This was the first time I had been entirely alone as a scientist. When I was working for Hal, he had the grand plan and knew what needed to be done. But here, once Richard Lambertsen was ashore, I had to feel my way through the opportunities, the risks, and the details of the job I had been asked to do. There were huge moments of self-doubt, but there really wasn't any option other than doing what I could do according to my best judgment. Years later, on a Woods Hole Oceanographic Institution (WHOI) ship collecting deep-sea fish for a toxicology project, I was talking through a request that had been sent to the ship to do something that I was not equipped to do. A good friend looked at me and said, "You can only do the very best that you can do, with what you have."

My second challenge was to acquire accurate data on animals that were killed efficiently, in that they were struck by the harpoon and immediately went into a dive, and the next thing I saw was a dead whale at the surface, with jaw open, flipper relaxed, and no obvious movement. I could not record the time when those changes occurred, as they did so underwater. For those cases, the postmortem pathology observed at the whaling station was critical information. In contrast, animals that were not killed efficiently tended to surface and breathe sometime after they had been shot, and in those cases, the behavioral observations were far more relevant.

After those many hours on watch, those significant challenges, and 18 dead fin whales, what did we learn? The details are recorded in our report to the International Whaling Commission.[3] This book is not a report, and I will not cover the most technical details, but some numbers from our study will help to describe what I saw and learned. When data from the whales were tabulated, the average time to death was 3 minutes, 42 seconds. The shortest estimated time was 0 seconds, and the longest, 16 minutes and 12 seconds.

The whales were numbered 24 to 42, which were cumulative catch numbers for the *Hvalur 6* in that season. Its hunting season had begun before our study.

July 13, 1983: Whale 30 was a 63-foot (19 m) male. When it was harpooned, it rolled onto its side and moved its flukes three times before submerging 22 seconds after being struck. There was blood in the circle of flat water where the whale had gone down, and the cable ran out slowly. Once the whale was hauled in, the jaw was slack, and blood was pouring from the blowhole. At postmortem examination, we saw that the point where the harpoon had entered was in the back, 6.6 feet (2 m) behind the blowholes; that it had then penetrated the skull, splitting the braincase in half; and that very little of the brain remained. The behavioral data suggested time to death of 22 seconds, and the postmortem data suggested instant death, 0 seconds. The latter seemed the most reasonable conclusion on the basis of the available data.

July 18, 1983: Whale 35 was a 64-foot (20 m) female. The grenade exploded before the harpoon hit the animal. The animal breathed 11 times before a second harpoon with line was fired 11 minutes later. It then breathed 13 more times before a killer harpoon was fired 14 minutes and 58 seconds after the first harpoon. It exhaled underwater at 16 minutes and 22 seconds, and the jaw was observed open at 18 minutes. The time to death was 16 minutes and 12 seconds. At postmortem examination, we found that the first harpoon was embedded in the blubber only, with no shrapnel. The second harpoon had penetrated the muscle over the chest, but not the chest cavity. There was extensive shrapnel damage in the muscle around that harpoon. The killer harpoon had penetrated the neck and caused massive bleeding in the brain. The estimated time to death was 14 minutes and 58 seconds.

These two cases cover the gamut of the observations that we made. At times, my behavioral observations were informative, and at other times they were not. The postmortem observations were substantially more detailed, and when we combined them with the timing of events observed at sea, we could make more accurate estimates. Again, more technical details are left to the report. For the purposes of this book, I will say that explosive harpoons kill whales within minutes, if not seconds.

We concluded our report by recognizing the great effort the gunners made to achieve an effective placement of the harpoon. It took, on average, five hours of tracking a whale before it was shot. The range at which the whales were shot varied from 15 to 100 feet (5 to 30 m). The whalers' goal of killing these animals efficiently was motivated by the desire not only to minimize their suffering, but also to achieve the greatest economy of time, effort, and materials. A quick kill was therefore an optimal outcome for multiple reasons, and the presence of an observer was not likely to have biased the data.

Spending three weeks aboard the *Hvalur 6* was a unique experience. I had been at sea for quite some time in a variety of vessels by that time in my life, but it was the first time I had been in an industrial marine environment, with such highly engineered equipment specialized for what was an inherently dangerous undertaking, given the high explosives, grenades, lines under heavy dynamic loads, large winches, and hugely powerful animals. I came away with extreme respect for the seamanship I had observed and the skilled human actions required to operate engineered systems to accomplish the task at hand. Inevitably, the major variable in the efficacy of each kill was the placement of the harpoon and its operation as designed—or not, as with the premature explosion of the first harpoon in the case of whale 35 above.

Fin whale carcass being hauled up at the Hvalfjörður whaling station, July 1983. The harpoon, which is visible at the top of the carcass, is poorly placed, not having penetrated the body cavity. A second harpoon, whose entry point can be seen between the visible harpoon and the left flipper, was required to kill this animal. Photo: Author.

It was important for me to gain the mutual respect of, and thereby a relationship with, the crew of the vessel. At the outset, none of them ventured to reveal that they knew how to speak English. Gradually, that changed as some trust developed. This experience stuck in my mind, and I would come to respect the same professionalism in mariners in the fishing boats that I worked on in later years in New England.

From Iceland, I moved on west, first to Labrador, then to New England. My strong sense was that the United States or Canada, at that time, was where I would be able to pursue a career in studying marine mammals. At least that was what my left brain was telling me. The other half was desperate to figure out a way to spend the rest of my life with Hannah.

Maybe my right brain had seen the travel to jobs in Iceland and Labrador as tickets to that end. I next spent two years at a veterinary hospital north of Boston, gaining membership in the American Veterinary Medical Association and ultimately a license to practice as a veterinarian in various New England states. Hannah was teaching music nearby. We got married the following year (1984), and by the summer of 1985, I had started a job at the Laboratory for Marine Animal Health, which was situated at the Marine Biological Laboratory in Woods Hole, on Cape Cod.

Hannah had been born and raised in a wonderful home just east of New Bedford, a forty-minute drive from Woods Hole. We moved there soon after we got married, so my search for a place to work was happily circumscribed by a commutable radius around that place. The following year, I enrolled as a PhD student in the Massachusetts Institute of Technology (MIT)–Woods Hole Oceanographic Institution (WHOI) Joint Program in Biological Oceanography. I proposed to study the interface between humans and marine animals in the Biology Department at WHOI, which is also in Woods Hole.

After leaving Iceland, I had a number of research directions that I was contemplating, more of the same kind of work, perhaps in Norway, involving the minke whale hunt there, pursuing some neuroanatomical questions that had arisen during the work in Iceland, especially where trauma to the spinal cord had resulted in behavioral changes that were not necessarily anticipated. However, when it came down to it, I was becoming most interested in the relationship between humans and marine animals. What humans did, in the case of an industrial whaling operation, was obviously devastating and lethal to their target animals. They were very efficient at what they did, but the idea that it was a necessary and desirable activity was certainly not obvious to me. This discovery intertwined in my mind with all the jumbled thoughts that had been rolling around since I had worked in Newfoundland and watched whales getting entangled, and heard of animals that had died from it. Those dead fin whales, with their explosive and rapid ends, joined the entangled humpback whales from Newfoundland.

Jon Lien and his crew were gradually developing disentanglement techniques. At the outset, in the late 1970s, many of the animals died despite disentanglement efforts, whereas by the mid-1980s, the majority were surviving, and—of equal importance to Jon—the amount of damaged and lost fishing gear was substantially reduced as the fishermen learned how to manage whales in their gear. I saw Jon in the early 1990s at a scientific meeting, and I remember catching up with him, and the wonderful twinkle in his eyes as he told me that more and more of the entangled humpbacks were being released alive.

However, despite my interest in the continuing problem of entanglement, I recognized that my interests lay as much in hands-on anatomical pathology as in working with live animals. Not finding a laboratory at WHOI that had access to dead marine mammals, I decided to step down the food chain to a fish-human relationships topic for my dissertation work. My research focused on the impacts on fish of chemical carcinogens in municipal sewage. Ten percent of the winter flounder in Boston Harbor had liver tumors. This finding was a significant factor in the court-mandated cleanup of the harbor. To collect the fish, I needed the help of a fisherman. After some inquiries within the community, I was introduced to Billy Crossen, a hardworking, bright entrepreneur who owned a wooden bottom dragger, the *Odessa*, and fished out of Gloucester, Massachusetts. Billy had a reputation for catching fish and for being a strong champion of his tightly knit community of dragger skippers. His disdain for government regulators was evident in his delight in catching skunks in his backyard and releasing them outside the regional fisheries management office in Gloucester. Billy took me fishing from 1988 until he died in 2007. We talked about where I needed to get fish, and he took me to where he knew the fish would be. He would leave port at 3:00 a.m. and not be docked till 9:00 p.m. He was an independent, freethinking soul who had a deep interest in the science and genuinely cared about the work we did together. At his wake in Gloucester, one of his friends, Mark Carroll, told me that Billy had told him to be sure to take care of my project, and that he should have all the related gear that was in Billy's basement. My relationship with Mark continues. These experiences have formed my perspective on commercial fishermen. I don't know any lobster or crab fishermen as well as I knew Billy, and know Mark, but the nature of their work is much the same—characterized by long hours, independence, and good people with deep knowledge of the resource that they harvest.

A bottom dragger pulls a funnel-shaped net, called a trawl, along the seafloor. The net is spread wide by "doors" that are rigged so that they pull the mouth of the net open. The bottom of the net is weighted, and the top has floats, to make a space for the fish to swim through into the tapered end of the net—the cod end. Once the net has been dragged for an hour or so, the steel tow cables attached to it are winched in and the contents of the net dumped on deck.

While we dragged for flounder, we would steer the net around the buoys marking lobster traps on the bottom. To catch lobsters, a suitable bait is placed in the trap, which is thrown over the side of the boat. The trap remains on the

A bottom dragger pulls a funnel-shaped net along the seafloor. The mouth of the net has a weighted footrope and a floating headrope. The mouth is spread open by two doors that are towed behind the vessel. Draggers pose greater risk to fish-eating whales than to right whales. © Woods Hole Oceanographic Institution, Natalie Renier, WHOI Creative.

bottom for a given amount of time before it is hauled to collect the harvest and rebait the trap. Inshore, a vertical line runs from every trap to a marker buoy at the surface. Farther offshore, a vertical line, or endline, is attached to the first and last of a string of traps—also called a trawl, just to be confusing. A groundline connects the traps in the trawl. Each fisherman's buoys are uniquely marked. Often, we would find abandoned traps and line in the *Odessa*'s trawl. They were hard to clear from the net, and choice words could be uttered.

We would also avoid gill nets. These curtains of net are set along the seafloor, with a weighted line to keep the bottom of the net down, and a floating line along the top of the net to keep the net vertical, so that fish will swim into the net. The net itself is a mesh of fine single filaments that fish cannot detect. Fish swim into the net and get entangled by the gills.

Anchors stretch the net along the bottom, and vertical lines run to surface floats that mark the location of each end of the net and enable it to be hauled to retrieve the fish.

Inevitable rivalries between different fishing sectors simmer, but absorbing stories from all of them over the years gave me a perspective on the industry as a whole that was hugely valuable to me, as well as some understanding of the conflicts among those who wanted to set traps for lobster, versus those who wanted to drag for bottom fish or scallops, versus those who wanted to set gill nets for bottom fish. The hours spent in the wheelhouse of the dragger listening to Billy—and now Mark, as the project is still ongoing—opining about the state of the fishing world have been critical to my understanding of that world.

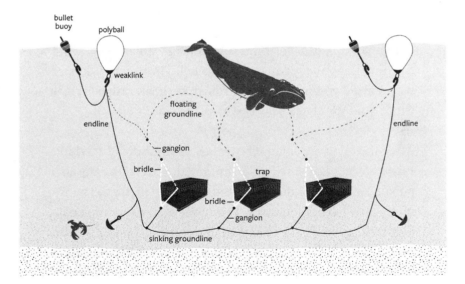

Characteristics of offshore New England lobster gear. The dashed line represents a traditional floating groundline. The solid line shows modifications made to lobster gear, such as weak links and sinking groundline, in the hope of preventing whale entanglements. © Woods Hole Oceanographic Institution, Natalie Renier, WHOI Creative.

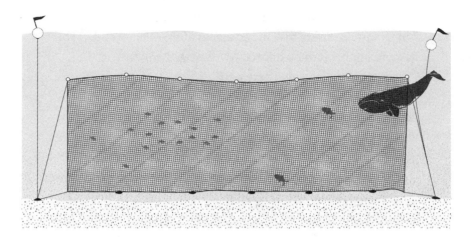

A bottom gill net, anchored at either end, with a weighted groundline and a floating headrope to intercept fish as they swim. The gear is retrieved by hauling on the endline that runs from each end of the net to a marker buoy on the surface. Right whales, humpbacks, and other whales can get entangled in gill nets. © Woods Hole Oceanographic Institution, Natalie Renier, WHOI Creative.

Operating a dragger requires endless vigilance to avoid catching traps and gill nets in the trawl. The dragger skipper has to avoid getting entangled in fishing gear, just as a right whale does. Neither is successful all the time.

After finishing my flounder studies as a graduate student, I carried on with that work, remained at WHOI, and also pursued a variety of topics related primarily to marine mammals and what humans do to them: their exposure to persistent contaminants; the effects of underwater sound stress on their diving physiology; the health status of live whales at sea, especially in the context of persistent fishing-gear entanglement; the management of live beached dolphins and whales. My interest in persistent contaminants led to a comparison of their impacts on different marine mammals that fed at different levels of the food chain. In marine food webs, the sun energizes microscopic oceanic plants (phytoplankton), which in turn are consumed by zooplankton, such as the copepods that right whales and fish such as herring then eat. Larger fish and some marine mammals eat herring. Then killer whales eat the bigger fish and marine mammals. At each step, persistent chemicals become more concentrated, to the point where killer whales, as top predators, carry concentrations of these pollutants high enough to affect their health and survival.[4] I wondered about the impacts of these chemicals on whales that are lower in the food chain. The right whale was an obvious species to compare with the top predators. Hence, I needed to find some samples from dead right whales to analyze.

A childhood friend of Hannah's, Amy Knowlton, had been working with Scott Kraus at the New England Aquarium, and she knew of my interest in such samples and told me about dead right whales when they appeared. However, she kept pressing me to help figure out why they were dying. The team at the aquarium had degrees in biology and conservation, but there was less capacity in terms of gross pathology and disease at that time. I rapidly realized that while contaminants in right whales could well be affecting their health, those effects would probably take generations to have an impact. Meanwhile, there were the obvious and immediate effects of vessel collision and entanglement trauma to be documented. Amy was right: we needed to know why right whales were dying. In this way, using my veterinary training to diagnose causes of death in floating and beached whale carcasses became a major preoccupation—an obsession, perhaps. I built long-term relationships with many collaborators, including the group at the National Oceanic and Atmospheric Administration (NOAA), National Marine Fisheries Service (NMFS or NOAA Fisheries) responsible for documenting and funding such things: the NMFS Marine Mammal Health and Stranding Response Program. Understanding why marine mammals died was essential to conserving them in the context of the US Marine Mammal Protection Act and Endangered Species Act. Essentially, I built the laboratory at WHOI that I was looking for as an incoming graduate student.

The privilege of being paid to be curious, through all of the above diversions and excursions into various aspects of the interface between humans and marine animals, left one persistent and overriding concern: humans have not been behaving well at all. We have been reproducing at an alarming rate, and our global demands for energy, seafood, and marine transportation of goods are having a very substantial impact on our marine environment, both climatically and in terms of wholesale reductions in biodiversity. The looming loss of the North Atlantic right whale, a species that has been killed by commercial activity for over a thousand years—directly by harpoons for centuries, and indirectly by accidental entanglement in rope or collision with vessels most recently—is a prime example of the latter. The market impetus for harvesting seafood and shipping commodities around the world is our consumptive culture. The reason why fishermen set traps for lobsters and crabs, and ships deliver manufactured goods and transport raw materials such as crude oil and its many derivatives, is simply because we as consumers create the demand for them. Thus, our very own consumer culture is the basis for this crisis.

I was yearning for an example of a human culture that could maintain a fundamentally sustainable relationship with its biological, physical, and chemical environment. In 2007, a gift to WHOI enabled its Arctic Research Initiative, through which I was granted funds to visit what was then called Barrow (now named Utqiaġvik), on the North Slope of Alaska, to study aspects of the Iñupiaq subsistence bow-head whale hunt. This ongoing native harvest has sustainably coexisted with bowhead whales for millennia, despite the worst attempts of commercial whalers from New Bedford and beyond to extirpate them between 1848 and the early 1900s.[5] Bathsheba Demuth, in her book *Floating Coast: An Environmental History of the Bering Coast*, traces the flow of energy as the driving force of human endeavor in that region. She describes how investors in the commercial whaling industry moved on to other sources of profit, such as the spring steel that replaced baleen in corsets and other essential goods, rendering it, and hence bowheads, commercially worthless.[6] The bowheads "survived because, to the world outside the strait, they ceased to have value at all." She also wrote of Harvard professor Alexander Agassiz, whose school of natural history, founded on an island 10 miles west of Woods Hole, inspired the founding of the Marine Biological Laboratory. Agassiz was quoted in the *Hawaii Star* as warning, in the mid-1880s, that commercial hunting would render whales extinct within fifty years.

The Iñupiat are now stoically adapting to the impacts of climate change, sea level rise, and sea ice loss. Perhaps the contrast between their sustainable subsistence harvest and the boom-and-bust Yankee whaler onslaught is a microcosm of the chasm between attitudes we see today: the "me, my business first" approach versus a more nuanced goal of equitable, thoughtful conservation of resources, enabling human and natural systems to be preserved, to be managed as a single unit, and to thrive over millennia.

My visit to Utqiaġvik gave me the perspective that I was searching for, one that could make sense of what we were doing to the North Atlantic right whale as an iconic example of how badly humans were messing up. I also wanted to bring the Iñupiaq skill set for taking the whales apart to the US east coast so we could examine dead right whales more efficiently. I did learn those skills, but I also learned so much more, about a sustainable whaling culture that could bring sanity to the east coast of North America by showing how it might evolve a sustainable coexistence with the North Atlantic right whale.

But I have got ahead of myself. I often do.

4

The Bowhead Is More than Food

Explosive harpoons, euthanasia drugs, the bows and pro- pellers of ships, and ropes and nets are all ways that humans kill whales today, whether by design or accident.

My time in Iceland provided the scientist in me with a baseline: if humans are going to kill a whale at sea, the quickest and most efficient tool is an explosive harpoon.

For well-meaning anti-whaling activists to say that commercial whaling with explosive harpoons is bad simplifies the reality of a far more complex series of relationships that humanity has evolved for killing whales with and without intent. There may be vegans who walk everywhere, homegrow all necessary food, clothing, warmth, and shelter, and never use goods or services generated in any part by consumption of goods moved by sea, but there are vanishingly few of them. The overwhelming majority of humans, myself included, especially in the developed world, eat seafood, buy commodities, and burn fuel transported by ships and barges. We are thus enabling human-induced morbidity and mortality of whales around the world.

. . .

But I had no sense of where I was headed when leaving Iceland. My agenda was complicated by a whirlwind of emotions: I needed a job, I wanted to share my life with Hannah, and my time in Iceland had created a kaleidoscope of intellectual and emotional ideas, concerns, excitements, and depressions.

I traveled from Iceland to Labrador, as I had committed to spend a further month with Hal on Hamilton Bank, east of Cartwright, using the behavioral research methods on fin whales that he had developed with humpback whales. Fin whales are larger and faster, which made his sailboat paradigm a bit more challenging. I had some critical experiences in that month. The first was an overnight in St. John's on the way to Labrador. I stayed with Lindy Weilgart, who, while approximating the innocent vegan described above, still burned fuel for heat, lived in a real house made of materials created in part from petrochemicals, and bought stuff. Her house was up against a rock bluff in the east end of the city, overlooking the harbor's fabled narrow entrance, aptly named "The Battery" in honor of the ballistic efforts of previous generations to defend themselves against would-be colonizers. Lindy confronted my willingness to lend my scientific credibility to the Icelandic whaling industry—the first time in my life that anyone had ever suggested the possibility that I had such credibility, which had certainly never been a belief I had held with any durability. It took me a while to focus on Lindy's concern, as my head was still processing the drama, the incredibly efficient and seriously heavy-duty engineering, the biology, and the trauma of the fin whale hunt, and recalling the professional whalers and seamen that I had come to know and like. I tried to explain that I felt my role had been to present an objective, factual record of the efficacy of explosive harpoons to inform the debate that was raging then, and still is today, among conservationists, animal welfare advocates, scientists, managers, and whalers. Facets of that debate include the ethics of hunting wildlife as a food source, the sustained value of live whales to whale-watch tourism versus the one-off income from a harvested whale, the unique nature of sentient mammals, and our ignorance of the difference in brain complexity between toothed and baleen whales. Perhaps the most important topic of all, given today's perspectives, is the ecosystem services rendered by whales: carbon capture, nutrient recycling from the bottom to the surface of the ocean, predator-prey relationships, and all the other things that they do that we do not yet understand.[1] In essence, this debate is about the rights and wrongs of messing with whales. That conversation with Lindy was hugely important for me, as our chats always are. It made me think deeply and drill down into what really mattered.

As I write this many years later, we are regularly exposed to remarkable media content. Recently, I watched a video of a human free diver pirouetting underwater in front of a stationary sperm whale. The whale then copied her spinning motion.[2] To put that interaction between a sperm whale and a human alongside the observation that sperm whales were a commodity, regularly killed by explosive harpoons and carved up at the Hvalur HF whaling station just a few years before I visited in 1983, is perhaps a reasonable perspective on the vastness of the chasm between how different humans view the sperm whale.

The evening, and next morning's breakfast, of post-Iceland decompression, debate, debriefing, and interrogation in St. John's had many facets. For many, myself included, the image of harpooning was appalling. It was painful to watch a beautiful, enormous wild animal that dives at will with grace and purpose having its skin, blubber, muscle, and chest organs pierced with a flying steel projectile weighing 154 pounds and tipped with a grenade designed to explode in the animal's chest, scattering lethal pieces of shrapnel around the chest, neck, and brain. However, having studied the basic science and practice of slaughterhouses at veterinary school, and the various ways farmed animals are killed, I was frankly surprised by the efficiency of the fin whale hunts that I had witnessed. I was still appalled, but there did seem to be a shade of gray that I had a hard time defining. Meanwhile, as a conservationist, I was, and remain, extremely concerned about our ability to truly know how many whales are in a population. Were the levels of harvest genuinely sustainable? Even for populations in which the vast majority of whales are known individuals, such as the North Atlantic right whale population, there are still major uncertainties as to how many remain alive.[3] My strong suspicion is that it is significantly fewer than we think. Furthermore, a previous generation of whale (whaling) scientists had needed substantial prodding by the supposedly less well-informed hippies of the 1960s, fueled by (among other things) recordings of humpback whale songs and publications such as the somewhat psychedelic writings of John Lillie, to wake up to the fact that the broad-scale commercial whaling of the time was utterly unsustainable.[4] In 2019, the Icelandic whaling effort claimed that its quota of 209 fin whales for the following five years was sustainable, but that is strongly contested.[5] The owner of the whaling company has defended its right to kill whales recently as strongly as he did to me decades ago.[6]

Lindy had put me on the spot by looking for a reasoned defense of the work I had done in Iceland. But the whole experience was still so raw that it was a struggle to rationalize it all so quickly. I was relieved to move on from my visit in St. John's to Goose Bay, Labrador, by plane, and then to Cartwright by floatplane, landing in thick fog, where I met up with Hal. He gave me a rather more nuanced and circumspect debriefing on my immersion in modern whaling, and set me to work on a larger, and less leaky, sailboat than the *Firenze*. We sailed through more ice and fog, and dodged gales by hiding in small harbors in off-lying islands, and then once again I got to watch the splendor of live fin and humpback whales without any concern that I would need to document one's demise. The difference between being aboard a small boat crewed by scientists, curious and engaged with these spectacular living animals, and being aboard a ship with professional seamen, making their livelihoods out of killing the same animals, was stark. I felt somewhat detached from both worlds at that point.

We spent a bad-weather day anchored at Grady Island. My journal recalls the sailboat rolling in a big swell in the marginal harbor on the southeast coast of the island. I went off in a small inflatable boat to look at the island's abandoned whaling station, then checked out the rocks in the tickle between the island and Little Grady Island. My journal records avoiding a significant risk of the inflatable capsizing. We were then able to feel our way into the tickle to get the sailboat out of the swell.

There are twenty-one abandoned whaling stations in Newfoundland and Labrador that were operational between 1898 and 1972.[7] In the past ten years I have visited some of them. At Grady, just a few winches and rusting steel tanks remained, which tends to be what one finds at such places, along with weathered whale bones, often in the surrounding shallow water. Like the dead whales that were processed

Abandoned winches, tanks, and other hardware at Hawke Harbor whaling station, Labrador, Canada. The amount of capital, and human endeavor, invested in extremely remote places by the whaling industry in the first half of the twentieth century is extraordinary. Photo: Author.

there, old whaling stations still evoke controversy. Some see them as precious industrial archaeological relics; others see them as decaying, depressing, rusting eyesores, full of hazardous materials that should be recycled, on land that should be returned to the wildlife and plants displaced by the structures when they were first built. Irrespective of their worth, they are monuments to the magnitude of the human effort and investment of capital that resulted in the global and nearly terminal exploitation of large whales in that era.

These relics of nineteenth-century industrial whaling can be found all over the world. There are seven abandoned whaling stations on the island of South Georgia alone.[8] In stark contrast to the appalling mess left behind there and elsewhere, a far more sustainable whaling effort has been ongoing in the Arctic.

Arctic native peoples have hunted bowhead whales for thousands of years. Anthropologists Sam Stoker and Igor Krupnik have described archaeological evidence of whale bones, including those of bowheads, being used in ceremonies at the Paleo-Aleut Anangula site in the eastern Aleutian Islands, dating from 6000 BC. And there is good evidence for active whaling in the past two millennia in the Sea of Okhotsk, on the Siberian coast, in the northern Bering Sea and Bering Strait, and from northern Alaska across Canada to Labrador, Greenland, and Spitzbergen.[9] Arctic peoples used the whale harvest for "food, fuel, and as a means to social status." The bowhead was "a slow and tempting target" with known migratory paths, supplying the hunters with oil, baleen, meat, and skin and blubber (known as muktuk). If European and American commercial whaling had not later exploited the same stocks for profit, it is doubtful that bowhead stocks would ever have risked depletion by aboriginal hunters using traditional techniques.[10] The precontact bowhead whale population had at least 15,000 animals and was probably much larger. The traditional harvest may have been 50 animals per year, or less than 1 percent of the population.[11]

Native human population numbers and densities were insufficient to generate unsustainable levels of yield, and religious and cultural traditions precluded harvest that exceeded need. Furthermore, the hunters' technology was not efficient enough to exploit the whales excessively, even if they had wanted to do so. Equally important, the lack of a profit motive, coupled with logistical limits on the number of whales that could be processed, meant that once a sufficient number of whales were harvested, motivation to continue the difficult and dangerous hunt rapidly diminished. But it did appear that active whaling enabled a greater community stability and size.

A major development in about AD 900 was the Thule culture, which spread from the northern Bering Strait west to Siberia and north and east across Canada to Greenland. It assimilated the earlier cultures, which resulted in a significant expansion of their whaling culture.[12] It was also the first to harness dogs for sled transportation. Thus, the bowhead whale was significant to the economy of the treeless Arctic, where the humans survived by hunting fish, birds, and land and sea mammals. Bowhead bones were also commonly found as structural members in Thule houses.

John Burns observed whaling on St. Lawrence Island in the Bering Sea, just south of the Bering Strait, in the 1960s and 1970s.[13] He wrote of the regrowth of ancient traditions that followed a steady increase in the bowhead population, the switch from dog teams to snow machines, and a combination of growing pride in and awareness of native traditions and values and improved economic status fueled by Alaska's oil boom on the mainland. The residents hunted for whales in open water using umiaks, undecked boats made from bearded seal skins. They used sails to approach whales quietly, then once the whale was dead, used groups of boats with outboard motors to tow the animal to shore. The towing boats were placed in order, with the boat that first struck the whale first, followed by boats captained by immediate kin, then members of the "same ranka (similar to a clan)," and then other boats in an order reflecting their role in the hunt. The bowhead harvest was shared with relatives and those involved in the killing, towing, and butchering. Burns described the whalers "praying and giving thanks, with a noisy and joyous ceremonial cutting, sharing and eating." He commented that prior to the addition of Yankee whaling tools, explosives, and outboard motors, the native whale harvest was a "very difficult undertaking that required the cooperation of several crews to kill, tow and butcher a bowhead."

The first known lower-latitude people to make contact with indigenous Arctic cultures were Vikings in the eleventh century. However, the Norse colonies were gone by 1500, and the native whaling continued. Basques then began one of the first industries in North America, if not the first, with their exploitation of the bowhead whale in the Strait of Belle Isle, between Newfoundland and Labrador, and off the coast of southern Labrador in the sixteenth century.[14] The fishery was at its peak in the 1550s and 1560s, but diminished substantially by 1590 as the bowhead population was decimated. However, the bowhead was not the first species to be exploited in this way. The Basques had begun with the bowhead's cousin species, the North Atlantic right whale, in the Bay of Biscay. Their earliest documented whaling activity was in 1059, and their latest around 1750.[15] Thus, North Atlantic right whales were no longer a prime target by the time the global onslaught against other whale species got underway. Sir Clements Markham, a nineteenth-century English geographer, described the Basque whalers as "dexterous whalefishers long before any other European people."[16] Other Europeans pursuing bowheads followed the Basques in the Strait of Belle Isle, and Dutch and Scottish whalers had had a significant impact on the bowheads of the Davis Strait by the 1700s. Other boom-and-bust commercial bowhead whale fisheries then ensued around the Arctic ice edge, often using native labor, equipping shore stations with European whale-

boats, and supplying trade goods.[17] American whalers in the
Pacific began to influence the Bering and Chukchi Sea bowhead populations, as well as native cultures, in 1848. They
hunted the whales initially from ships offshore, and later
using shore stations. They also affected native people by significantly reducing their other food sources, such as walrus.
The introduction of more efficient whaling technology
and a major influx of American ships made such serious
inroads into the bowhead population that commercial bowhead whaling ceased by 1910.[18] Since then, native whaling
in northwestern Alaska has variably continued to this day,
and the bowhead population has recovered well. In 2011, the
Bering-Chukchi-Beaufort Seas population was estimated to
be 17,000–19,000 animals and growing at 3.7 percent annually.[19]

The status of the native communities that still harvest
bowhead whales in northern Alaska is also good. They are
part of the mainstream consumer economy, yet they have
maintained a strong spiritual and cultural connection to the
bowhead hunt. Ironically, their ability to sustain their place in
the consumer economy does depend on seasonal resupply by
barges, which pose a risk for whale trauma.

My exposure to that way of life has been brief, but intense.
In October 2007, I flew to Utqiaġvik, Alaska. Utqiaġvik is
500 miles closer to Seattle than Seattle is to Boston, but cul-
turally, it feels much farther. Stepping off the plane into the
small terminal, I saw family reunions, warmth, purpose, and
vigor. Utqiaġvik is one of the oldest and coldest permanent
settlements in the United States, with nearby archaeological
remains going back four thousand years.[20] Driving to the
Naval Arctic Research Lab, a Nissen hut where I would stay
for the next week, I was struck by how temporary the modern
buildings appeared in a place with such history. What was the

Harvested bowhead whale, Utqiaġvik, Alaska, October 8, 2007. The whale is lying on its right side, with the baleen growing out of the roof of the mouth to the left, overlying the mostly white chin to the right. Photo: Author.

basis of the long survival of the native residents' forebears in the absence of the current modern assets those residents now use—such as fossil fuels, modern building materials, vehicles, ATVs, and air transport—during the millennia of living and surviving in a complex coexistence with nature? It was whaling, sealing, fishing, and dog teams. They used furs, skins, baleen, and much else before the advent of diverse petrochemical products produced by an industry that had itself hugely altered their ongoing lives, given the Alaskan North Slope's major role in that industry.

My motivation for visiting the community was to learn how they took bowhead whales apart manually and rapidly. On the US east coast, we often find dead whales on the beach that we struggle to disassemble to figure out why they died, and I knew that there was a deep resource of skill and multigenerational experience on the Alaskan North Slope that could help me get better at it. I was also keen to look at some details of bowhead bone structure to compare and contrast it with changes I had observed in right whale bones on the east coast and bones from bowheads harvested by Basques in the Strait of Belle Isle in the early 1500s.[21] I had read, and talked with colleagues, about the spring and fall bowhead hunt in Alaska. In the spring, the hunters use umiaks, which are light and can be taken to the edge of open leads in the ice. The hunters set up camp at the ice edge and wait there for whales to pass by. When a whale is killed, many villagers haul on heavy line leading through multiple pulleys to get it out of the water and onto the ice, where it is butchered. In the fall, the whalers use outboard-powered boats to hunt in open water and to tow whales to the beach. The meat is preserved in underground ice cellars dug into the permafrost.

In Utqiaġvik, I began to feel as though I had arrived at a place where I could begin to understand the insights into whale trauma that I had been puzzling over for so long. Here was a community that understood a whale species as well as any group of humans on earth. They had intertwined their lives with those of the species for thousands of years, and they had done so sustainably. The whales were growing in numbers. The native communities that had whaling crews had fewer social issues than those that did not. There was an intense sense of pride in their relationship with the whales. This interaction was so different from the confused and con-flicted ones I had been watching on the east coast, where whales were at once both a problem and an attraction. They got in the way of ships and fishing gear, imposing economic and regulatory costs on those trying to make a living from the coastal ocean. But they were also the basis of a burgeoning worldwide economy of whale-watching tourism.

In trying to understand the heart of the Iñupiat, I spoke with Craig George, who had moved to Utqiaġvik in 1977 and had been a wildlife biologist at the Department of Wildlife Management of the North Slope Borough since 1982. I had first met Craig in Boston at a 2003 workshop, where we compared and contrasted North Atlantic right and bowhead whales and considered how research on and management of the two species could be synergistic. He has described his role as an "all-consuming job of working the border between whalers and science." It was Craig who had given me my first window into the complexity of the relationship between the Iñupiat and the bowhead. Before that, the bowhead hunt was an anachronistic anomaly to me. In a recent email, I asked him how Iñupiaq whalers view themselves and their relationship with bowheads. He responded in an email on March 16, 2019:

Your question is a tough one. I can't say I fully understand Inuit whaling. What I've seen is a complicated mix of: selfless sharing, great humility & great pride, spirituality, survival, reverence and detachment for the animal, incredible bravery, great elation mixed when a whale is taken and too often terrible tragedy—and that's the short list. I try to think of the 1000s of years of living along the arctic coast in a 10 × 12 ft sod house through the dark months with 10 or more people. Women sewing skins preparing for the hunt, kids playing. The qargi, or men's house, were not much bigger; there the men prepared for whaling, skinning boats, making tools, selecting crews, deciding where & when to hunt, cutting trail, always believing the whale knew what you were thinking, and how you've conducted yourself. I wonder if selection pressure & kin-selection are a big part of the mix. Remember many decisions were made by the umaliqs (boat captains) by consensus and shamans were important too. A complex, sophisticated but very different society than ours. Even after forty years I can't quite wrap my head around it. The whale hunters essentially say—"you just have to live it, don't try to analyze it."

There is a wonderful collection of oral interviews with many Iñupiaq whalers from Utqiaġvik and other whaling communities in Alaska, as well as with Craig George, archived by the University of Alaska Fairbanks Oral History Program, several of which are available online through the Northern Alaska Sea Ice Project Jukebox, which recorded traditional, inherited knowledge about sea ice.[22] Understanding the behavior of sea ice is critical to the safety and success of subsistence whaling, given that it is the constantly changing platform that the native whalers use to hunt from during the spring migration. In a book about Inuit sea ice knowledge and use written by Matthew Druckenmiller, Craig George, and others, Arctic ice is described as "one of the more complex, ephemeral terrains on earth."[23] Wind direction and speed, air and water temperature, changing currents, and water levels all contribute to the massive forces involved in shaping the ice conditions of the moment.

In two 2007 interviews with Karen Brewster, Craig George described how the scientists in Utqiaġvik came to understand the unique value of native knowledge handed down for many, many generations.[24] His work included bowhead abundance estimates, descriptive whale research (basic anatomy and physiology), and compiling native knowledge of whales, sea ice, and Arctic life. Recognition of the value of what is now referred to as Traditional Ecological Knowledge was an approach started by his supervisor, Tom Albert.

George quoted Albert as saying, "The bowhead fell off the moon in terms of science," meaning that there was a lot of knowledge about it in the native community, but not so much in the scientific community. George had been taught in university not to trust the knowledge of stakeholders such as cattlemen and hunters—that science knew better in terms of understanding and managing wildlife. He described Albert's respect for native knowledge in Utqiaġvik as a paradigm shift. The Iñupiaq skill in camping and surviving on sea ice was essential to the team's effort, as collecting Arctic field research data safely took 90 percent logistics and 10 percent science. Albert had been trained as a large animal veterinarian. On coming to Utqiaġvik, he quickly built relationships with the native community and listened to them when they told him that scientists were grossly underestimating the number of whales. His veterinary training had avoided the bias of the primacy of science that George had had instilled as an undergraduate studying wildlife management. Albert's mentor in large animal practice had taught him to take a history: "Listen to the farmer, the owner: ask them what is going on with their animal. The history: listen to those guys, they know their animals." George had to depend on the local people to safely set up an ice camp. He could not count bowheads until he understood ice types, weather, whale behavior, and migration timing. He had to learn from the natives. He had learned risk management through climbing Denali, and he transferred that knowledge to the sea ice. In addition to cold and bears, the big risks are getting stranded on a broken-off ice floe and "getting smacked by the pack ice." The older whaling captains walked him through the basics. He learned that the whales are now coming earlier: in late March, rather than sometime in April. The hunters told the scientists where to set up their whale counting perch. He described how the ice had cracked up, but left them a peninsula to retreat along. He understood that ice was moved more by current than by wind. He repeated what the elders always said:

> You've got to watch ice develop, where the iiguaqs [floating ice sheets offshore that join the shore-fast ice and may become anchored pressure ridges] are and get a sense for how long it's been in place. Where the big ridges are, where they're not.[25]

He described the time in the 1970s when the ice never went out in summer, and the lack of sunlight for the marine algae, on which the bowhead's food needed to graze, in turn made it hard for the bowheads to find enough food. He suggested that the reason why bowheads have extremely thick blubber is to survive such lean times. He also talked about how the native hunters are adjusting to the decreasing amount of ice, earlier spring whaling, and later fall whaling.

> The hunters needed us, we needed them. And we developed a really good partnership and it should be a model for other resource users. Don't fight with each other. Work together. Maintaining a reverent whale hunt here is an important thing for a lot of reasons. And maybe some wouldn't agree, but as long as it's done in a sustainable, reverent manner, it's a huge contribution to the community and to science. It's unbelievable what we've learned about bowheads by being able to watch them in the spring, and then as Tom would say, "look inside 'em a couple times a year."[26]

This root system of traditional and fundamental knowledge is evident throughout the interviews curated in the Northern Alaska Sea Ice Project Jukebox. I quote from three of them here to give a tiny window into the breadth and depth of the natives' understanding of their Arctic environment and how it is sustained within their culture and heritage. Lewis Brower described how they are able to assimilate and communicate an understanding of the effects of weather and currents:

How do we pass that on? I think the first thing that you gotta pass on is to make sure that you know when you're out there, what you're looking at and how to read it. We try to pass that information on as we grow. It's weather patterns that we look at, which includes watching the currents in the ocean, watching the wind, watching the ice flow on the other side that's moving. You know that there was a recent storm in the Bering Sea and all the water's being pushed into the Arctic Ocean from the Bering Sea, which actually will lift up the ice when we're out there. . . . We still have to learn to read it, and it's all about being out in the conditions that we have and being with it. Kind of like being as one. When you're out there, you learn—you learn real quickly what you listen to. You know, if you hear the creaking in the ice, if you hear how the wind howls, if you hear the waves, you know, what do you do with all of that? How do you put two and two together so that it's safe for everybody? You're not going to learn that in one day. You're not going to learn it in a month. You're going to try to understand it in that month, and then as the years that you hunt—they start connecting together.[27]

Billy Adams spoke of the role of storytelling in passing down knowledge:

I learned a lot of my hunting through stories—listening and putting it into my memory, my brains. What did you know—

like a movie into my brains. So, you know, every time I go out I really enjoyed who I learned from. There was a lot of other captains who I really learned from, too. You know, my Uncle Robert Aiken and then Tom Brower, you know, I'd go to his store when he had his store. You know, after school I'd go hang out with him when he liked to tell me stories for a couple hours then I'd go home and do my homework.[28]

Jacob Adams, when asked if climate change, thin ice, and warm weather are the new normal and what they mean for whaling, told of the community's inherent resilience:

If it gets to that point, we'll just have to figure out ways to deal with that issue . . . our culture has always been faced with challenges, so how do we get the animals that we use for subsistence. Our ancestors didn't survive by saying what do we do now. We just have to look for ways to solve those challenges and work around them.[29]

When I visited Utqiaġvik in October 2007, I watched the whalers butcher six bowhead whales, learning what made them so adept at taking the animals apart, and also collected some samples for the bone research that I was undertaking at the time. I also visited the Iñupiaq Heritage Center.[30] In addition to the many artifacts on display, there are clues to the closeness of the Iñupiat to the bowhead whale. There is a quote in an exhibit attributed to Herman Rexford, of Kaktovik, in 1991: "There is one teaching where a person is told to be careful about being too outspoken, especially where it comes to talking about a whale, for it hears all that is being said about it. The whale knows when people are talking about it." Another exhibit lists Iñupiaq values: love and respect

for their elders and one another, respect for nature, knowledge of family kinship and roles, sharing, knowledge of lan-
guage, cooperation, a sense of humor, knowledge of hunting
traditions, compassion, humility, avoidance of conflict, and
spirituality. There is also a concept of the whale being a light
burning in an oil lamp. The flame is tended by a young girl,
who steps away to breathe only when the whale surfaces. If
the light is extinguished, the girl dies. The spirit of the whale
is given to the women. The whale's spirit tells other whales of
its kind treatment by the women and convinces those whales
to give themselves to the whalers the next year.

According to another exhibit, there are over forty whaling
crews in Utqiaġvik. The captain of each crew has a unique
flag and whaling tools passed down through the generations
along with the necessary traditional knowledge. The exhibit
illustrates how umiak frames are made and covered with
the necessary five to six bearded seal skins, which are sewn
together with braided caribou sinew and waterproofed with
seal blubber. Most communities whale only in the spring,
from the ice edge in umiaks, but in the fall in Utqiaġvik,
whales come close to the shore, so fall whaling occurs there
as well.

The following year, in February 2008, I was able to return
to Alaska. My first stop was at the Alaska SeaLife Center in
Seward to give a lecture at a meeting of the Alaska Marine
Mammal Stranding Network on the protocols for large whale
necropsy (the animal equivalent of an autopsy) that we had
developed on the US east coast for determining how stranded
whales had died. The crowd was receptive, but the logistics of
our protocol seemed rather impractical for Alaska, given the
remoteness of the places where many of their cases stranded
and the problem of bears scavenging the carcasses. I was

able to share the methods we had developed, and my experience in diagnosing deaths from vessel collision and fishing-gear entanglement, but it struck me as bizarre to be teaching whale disassembly techniques I had refined in Utqiaġvik the previous fall, when there were far more skilled folks in the same, albeit huge, state.

I then returned to Utqiaġvik for the Alaska Eskimo Whaling Commission whalers' convention. It was striking to compare the efficacy of native whalers in protecting bowhead whale habitat from oil and gas exploration threats with the efforts of the North Atlantic Right Whale Consortium. The consortium consists of right whale biologists, conservationists, managers, interested nongovernmental organizations, and shipping and fishing industry stakeholders who have come together with the intent of sharing knowledge about North Atlantic right whales to ensure their conservation and growth.[31] In Utqiaġvik, I saw a group of native whalers having far greater success at conserving their bowhead resources through habitat protection than we were having in trying to keep the North Atlantic right whale from extinction.

In this chapter, I have attempted to lay out my brief and superficial perspective on native subsistence whale hunting. A far more elegant summary has been written by George Noongwook, chairman of the Alaska Eskimo Whaling Commission:

> To our people, the bowhead is more than food. It keeps our families together. It keeps our children in school. It allows our elders to pass generational knowledge to our youth. It teaches us patience and perseverance. It teaches us generosity. It strengthens our community. It provides wisdom and insight. It gives us hope. It is our way of life. The spirit of the whale lives within each of us.[32]

Thus, although the subsistence whale hunt in Alaska has evolved technically, it remains a balanced, deep-rooted cultural wellspring. Yet many whale conservationists and anti-whaling activists argue that the very idea of killing a whale for food today is fundamentally unthinkable. All it took for me to make up my mind was to walk into a grocery store in Utqiaġvik: I was shocked to see the price of a box of cereal or a gallon of milk. The necessity of harvesting local food sources such as whales and seals was obvious.

In the past two decades, I have been increasingly exposed to dead whales on the beaches, and floating offshore, on the east coasts of the United States and Canada, most of them struck by vessels or entangled in fishing gear. I have also spent time consulting with colleagues as to whether, and if so how, we could euthanize large whales with a terminal condition, such as intractable chronic fishing-gear entanglement. Thus, in Alaska, I was beginning to find a place where killing whales made some kind of sense, as compared with the senseless killing off the east coast, where accidental trauma from ships and fishing gear results in a decomposing mess of dead whale that does no good for any involved: neither the human steering the ship or setting the fishing gear, nor the whale, especially if the trauma takes months to finish the job.

5

Whaling by Accident

I spent much of spring 2020 working on this book at home with Hannah, hiding from the novel coronavirus, she teaching elementary school band remotely from the kitchen, and me upstairs pecking away at this book. After about four secluded months of that, we agreed that we needed to have a change of scenery, but safely. Our sailboat took us to the cove inside of Long Point in Provincetown, at the northern tip of Cape Cod. We anchored there for four days, contemplated the state of the world, and rowed ashore to hike the sandy beaches. We watched and listened to the gray seals resting in the shallow water a hundred yards upwind, and the coyotes in the dunes. Both animals' distinct, eerie calls were timeless and reassuring. Common and least terns fed on the swarms of baitfish around the boat, flying into and out of the fog on a windless dawn. Their harsh, shrill calls were mixed with sounds of watery swirls of larger fish rising in the bait swarms. Things we could have seen if we were there a thousand, or ten thousand, years ago. We talked of whale, dolphin, and seal cases I had worked on in the previous decades in that town. The minke whale that had died in front of a restaurant at the foot of the Provincetown pier. The young humpback whale that had been drowned in lobster gear as the tide came up off the western shore of Cape Cod Bay and then floated across the bay to wash up on the beach at Race Point, Provincetown. The well-known local fin whale that had been hit by a ship and washed ashore on the south side of Long Point a long stone's throw from where we were anchored. The many young seals found with pieces of gill net around their necks. These seal cases are something of an enigma. Presumably they push their heads through the mesh of a net, but it's unclear whether they tear the net to free themselves, or whether their weight in air tears the net as it is hauled out of the water, or whether they are cut free by fishermen as they reach the boat's hauler. Whatever the case, they die months later when the pieces of net constrict their throats as they grow.[1]

While in Provincetown, Hannah and I worked with Lisa Sette, from the Center for Coastal Studies, to fly a drone over the seals to count them, learning that for every seal on the ocean surface, there were three resting on the bottom. We also flew the drone, equipped with an infrared camera, over some fin whales to learn more about the heat signature of their blows as part of a project we are developing to enhance detection of whales by ships.

Back in 1990s, I did not know the endless sandy beaches of Cape Cod, and their wildness, so well, but I did understand that it was important to accurately diagnose the reasons why marine mammals we found there were in poor health or dead—not only for the sake of species conservation, but also for the health and welfare of individual animals. It was obvious that trauma from vessels and ropes was a very significant concern, but without the best diagnostic information possible, we could not address the issues efficiently. The challenge was to know where and when the trauma occurred, and what caused it, so that efficient remedies could be established. For animals that remained anchored by, and drowned in, marked fishing gear of a known type, it was relatively easy to add their cases to the weight of evidence showing that something was needed to reduce that specific risk. Not so much for a whale that had been swimming for months, trailing rope that slowly constricted its vital organs, before it washed up on a beach. In those cases, we did not know where the entanglement occurred, or when, or the fishery involved, unless there were identifying marks on the line, or some characteristic configuration of knots and rigging. The knowledge that rope of some kind, from some source, was at fault was not particularly useful. It was also essential to know whether the evidence in hand was the cause of the animal's ill health or demise, or incidental to those endpoints. Over the following years, we slowly built up a series of cases, with detailed documentation, of carcasses that we had examined to better understand the causes and effects of trauma from rope and vessels.

On March 9, 1996, I was in Provincetown, visiting the Center for Coastal Studies. News came in that a dead North Atlantic right whale had floated ashore in Wellfleet. We drove back down the asphalt spine of Cape Cod: Route 6, which stretches from Provincetown to the Sagamore Bridge over the east end of the Cape Cod Canal, and then all the way to Bishop, California. The highway runs "uncertainly from nowhere to nowhere, scarcely to be followed from one end to the other, except by some devoted eccentric."[2] For me, it has been my path to and from countless marine mammals found on the beach, "strandings" that I have worked on, with many colleagues, while living in that area for the past thirty-five years. Since 1998, I have worked with the Cape Cod Stranding Network, which became part of the International

Fund for Animal Welfare in 2005. Most of the strandings occur on the western, Cape Cod Bay side of the Outer Cape peninsula, landing on one of the many beaches, or most especially within Wellfleet Harbor, which is enclosed by a long sandspit running south to Jeremy Point and wrapping around a complex series of creeks and sandbars, within which all manner of marine mammals go to die, or strand alive. Jeremy Point is the barb of the Cape Cod hook, of which Provincetown is the sharp, recurving point. However, this whale was on the eastern side of the peninsula—on the "backside," as it is known locally—on a beach that is wide open to the North Atlantic Ocean.

The outer beach runs 40 miles from Race Point in Provincetown south to Chatham and on down to the southern tip of Monomoy Island. Sand, often with sandy cliffs, is occasionally interrupted by a series of inlets, shallow channels, and sandbars. We have seen a few dead whales on that coast over the years. Longshore currents, surf, and shifting sand make those cases hard to manage and examine. In this instance, the whale had been reported by the National Park Service: the majority of the outer beach area is part of the Cape Cod National Seashore. The whale was on Newcomb Hollow Beach, in Wellfleet.

I first examined this animal with staff from the Center for Coastal Studies. It was lying near the low-tide mark about a mile north of the Newcomb Hollow parking lot. We examined the carcass by flashlight. There was rope through the mouth and around the tail. The wrap around the tail was entangled in two of the bones (vertebrae) that form the animal's spine, which were protruding from under the animal's body. The rope was attached to the squashed remains of a lobster trap that had a Canadian tag indicating that it was set in the Bay of Fundy. We removed the tag in case the carcass floated off on

Remains of dead North Atlantic right whale #2220, March 9, 1996, Newcomb
Hollow Beach, Wellfleet, Massachusetts. The black skin covers the blubber
coat, head, and tail. The muscles, viscera, and skeleton are largely absent. The
tail is to the left, with an entangling rope running to a compressed lobster trap.
The head is to the right. The jawbones are pushing the throat blubber into two
parallel curves. Photo: Author. Permit: NMFS authorization to New England
Aquarium.

the tide. The animal, a male, was 44 feet (13.4 m) from its tail
to its head. Later that night, anchors were used to secure it.

The whale carcass was best described as a giant, empty,
black tube of toothpaste, the tube being the blubber coat with
the black skin still attached. There were some bones inside
the tube, but vertebrae and ribs were squeezed out of the
tube's mouth and scattered in the surf as the animal rolled
down the beach over the few days that we took to examine
it. The combination of the deflated state of the carcass with
the skin still attached to the blubber was puzzling. With the
amount of internal decay we observed, we had expected
the skin to be long gone. Only years later did we learn that a
freshly dead right whale that has washed ashore on a high-
energy beach, pounded by surf, can roll for miles down the
beach. The surf macerates and liquefies muscles and viscera,

and its relentless pressure squeezes the bones and liquefied soft tissues out of the mouth, but if the whale is freshly dead, the skin can remain attached to the blubber.

Over the next few days, a group of us, including Amy Knowlton from the New England Aquarium, examined the animal. It was remarkable how much we could learn, despite the fact that 80 percent of the whale's body was missing. Amy was able to match some of the markings to an individual in the aquarium's ID catalog. The whale was a 4-year-old male, #2220, first sighted alive in Florida on January 12, 1992, and last seen alive on October 3, 1995, six months earlier, in the Bay of Fundy, Canada.

The next day, we further examined the whale, which had rolled about another half mile down the beach. One of my goals was to measure the thickness of a live whale's back fat. This might seem a strange priority, but I thought it might give us a clue to a problem I had been discussing with Amy. We knew that there were not many North Atlantic right whales left, and that they had trouble reproducing—that what we call "calving success" was poor. We needed to develop a way to measure the extent to which the health of adult female right whales was being affected by food availability and sub-lethal trauma from rope entanglement—a significant piece of the income and expense balance, as discussed in the preface to this book. The whales should be having calves off Florida every 3 years, but the intervals between calves were often much longer, maybe every 6 or 7 years on average, and recently up to 12 years. As a veterinarian, I had been trained to measure cattle back fat, using an ultrasound machine, as an indication of reproductive health. Amy and I wondered if we might do the same with whales—despite their blubber being much thicker than a cow's fat layer. In hindsight, this was the beginning for me of decades of thinking around the problem of right whale health—about how a whale's body condition reflects its health, and the role a better understanding of that relationship could play in managing the species to recovery. When he heard my proposal, Stormy Mayo, from the Center for Coastal Studies, offered, with a half-smile, to put me in the back of his pickup truck with the ultrasound probe on a long pole and drive past the whale on the beach to replicate the experience of trying to measure the blubber thickness of a live whale at sea from a moving boat. He was not the only skeptic about my plans.

We all talked about how to manage the carcass and decided to pull the animal up the beach as the rising tide floated it. I had ideas of towing it to a boat ramp, but it was a long way to anything suitable. In any event, an anchor line was attached to the peduncle, and National Park Service personnel winched the animal up the beach as the tide floated it.

The next day, five of us undertook a partial necropsy. The carcass was indeed a stinky, skin-covered blubber bag full of bones and strands of rotten connective tissue. Rolling in the surf had liquefied the soft tissues, such as muscle, heart, and viscera. Typically, the last to go is the largest vessel coming out of the heart, the aorta. But in this case, the soft tissues had all gone. Bones were haphazardly jumbled within the body cavity. The animal was "deflated," having a height above the beach of only 2 to 3 feet, about a fifth the height of a recently found right whale I had seen in Georgia, and a far cry from the freshly killed whales I had seen being butchered in Iceland thirteen years previously.

This was the first time I had seriously waded into the semi-liquid remains of a dead whale. We were unable to fully examine the skull bones, although fractures were suspected, as the rising tide buried their secrets. But we did examine and collect some of the bones. We also found the tongue, shaped like the front end of a Volkswagen Beetle and about the same size, and quite resistant to decomposition, being fibrous and oily. Numerous old and fresh rope scars were present around the upper part of the head, including a deep incision across the roof of the whale's mouth and another around the peduncle. The peduncle is a very common place for rope entanglement because the broad flukes operate much like a cleat, commonly used to hitch boats to a dock. Wrapping rope around a cleat uses friction to keep the rope and boat in place. That's good for a boat, but bad for a whale.

In addition to large knives, shovels are valuable tools at a beach necropsy of a large whale. We dug out the area around the vertebrae and the lobster trap by removing the underlying sand. This revealed the back end of a laceration that began about 3 feet (1 m) in front of the peduncle. It appeared that the tail end of the spine was detached from the rest of it and was protruding though a cut that penetrated the full thickness of the blubber. The front third of the cut consisted of three or four diagonal slices, apparently caused by a ship's propeller. At this point, the tide made us quit for the day.

The next day, Amy and I arrived to find the whale's body loose in the surf again. The lines that tied it to a truck and a backhoe had broken as the receding tide pulled the carcass back into the ocean. We finally caught it on the beach and collected some remains of the skull. We also found signs of infection under the blubber to the left of the blowhole and in front of the right flipper.

After a series of attempts to learn what we could from this animal as the surf and tide rolled it up, down, and along the beach, we knew that the whale had polypropylene line, the kind widely used at that time for lobster fishing, around its mouth and tail. It had a large incision, with some propeller cuts, from the left kidney area to the tail. It was unclear whether these cuts penetrated below the blubber layer. The left upper jawbone appeared to protrude from the roof of the mouth. The animal also had infections and an abscess below the blubber.

Before determining whether a vessel killed an animal when propeller cuts are present, it is important to establish if the cuts were made before or at the time of death, or afterward. A boat can strike a whale after it has died. If the tissues are fresh enough, we might see evidence of bruising and bleeding, both of which require a beating heart: this is good evidence of the trauma preceding, and potentially causing, the death of the animal. However, this animal was too decomposed for us to decide.

We considered the following scenarios: a ship had struck and killed the animal—given the apparent jawbone displacement and the lacerations toward the tail; it had died of infection; or the rope entanglement reduced its feeding and swimming efficiency. However, its advanced state of decomposition made any firm conclusions difficult. Perhaps all three factors contributed to its death. We did, however, secure some blubber samples for the ultrasound project.

Thus, despite all the hard work we did, with a huge amount of patient support from the National Park Service rangers, over multiple days, we knew a lot about the animal, but were unable to finish our job: to diagnose the cause of death. Over the succeeding years, I have learned, from and with others, how best to interpret these scenes in less time and with more efficiency. This case is also instructive in terms of the trauma that is so commonly seen in dead whales washed up in industrially active parts of the oceans of the world: namely, signs of both fishing-gear entanglement and vessel collision. I was beginning to understand what others had been describing for some years.

My work on these cases, to this day, creates a sense of numbness in my head: fear of missing the critical observation, of not knowing the most important thing to do in the limited time available; worry that it won't be until it's too late that we realize what we should have done better. I don't think I have ever been able to drive home from a large whale necropsy without replaying the event in my mind and thinking, "Why didn't we sample that bone, make that sketch, or take that photograph?"

You may wonder about my urgency: the animal had already died, so it was not as if more time would have allowed us to save its life. But a complete diagnosis might have helped us to save others. This case was the sixth North Atlantic right whale death to be recorded in the first three months of that year. I knew that without diagnoses, there could be no hope of solutions that would prevent further deaths. I felt the weight of our responsibility to provide accurate diagnoses so that stakeholders, such as the shipping and fishing industries, would not be asked to make unnecessary, expensive changes in the way they worked.

I have learned a lot since that Newcomb Hollow case. For one thing, I learned the key elements for a successful necropsy. One is a disposal plan that makes the manager of the beach keen to see you do your work as a step toward disassembling their unexpected nightmare of a dead whale, often where folks normally spread recliners, beach umbrellas, and suntan lotion. Another is a relatively small team of people who know how to use (and sharpen) a knife efficiently, along with a good photographer, recorder, and sample manager; a person who looks out for the safety of the entire scene; and someone to manage bystanders and news media. Another is a means to get the whale's body above the high-tide mark, which in turn calls for the biggest excavator available, or perhaps two or three. The biggest asset is the operator of the machinery. Diesel and the power of hydraulics are your friend.

In terms of bettering my own understanding of whale mortality, I seriously considered further training as a veterinary pathologist. Such training would have required residency at a suitable laboratory and passing the examinations for board certification with the American College of Veterinary Pathologists. I was the father of four young sons at the time, all attending the school where my wife was happily embedded in her teaching career. To pass those board exams, I would need to spend significant time away. I decided to stay put and learn the gross pathology I needed on the beach. I know today that I put family before job in that decision, but such are the decisions one makes. However, I have had the privilege of working with a good number of board-certified veterinary pathologists in the intervening years, and I am extraordinarily grateful for their knowledge, understanding, experience, and patience.

Revisiting our findings for the whale in Wellfleet today, I would certainly lean toward the vessel strike as having occurred before the animal died, which then resulted in its rapid demise. I also learned from that case that indeed, getting the animal above the high-tide mark as quickly as possible is critical if a good examination is to be undertaken. That would have required a much larger machine than was available from the National Park Service, and therefore money we did not have at the time.

All of those lessons were quite granular: at the level of logistics, process, and interpretation of data from an individual animal. However, as the years went by, many colleagues and I worked on other dead whales—right, humpback, fin, minke, sperm, and blue—hit by vessels or entangled in rope. We began to build a much broader perspective on these two pervasive impacts humans were having on large whale morbidity and mortality.[3] The data were hard to ignore. Indeed, when the federal government was considering enacting regional and seasonal speed restrictions in 2007 to minimize vessel trauma, there were stories of the actual necropsy reports being reviewed by the vice president at the time, Dick Cheney, in the White House. It felt good to have the detailed reports of trauma that I had sweated over become part of a high-level assessment of what could and should be done to manage the problems.

Meanwhile, back in Woods Hole, Carolyn Miller and I looked at the blubber samples in the lab to test the ultrasound device. Carolyn had been in my lab for a while, first as an undergraduate looking at the responses of marine mammals to persistent chemicals. Later, she took up the back fat project as a PhD dissertation. For quite a while, we were unsuccessful in convincing ourselves that the system was going to work with whale blubber. However, we did get echoes on the screen of the machine. It was only after we realized that the samples we had collected had come from a part of the body where the interface between the blubber and the muscle is made up of two discrete layers of fibrous tissues, that we recognized that area as a double echo on the screen. We then knew that the tool actually had promise.

It is these moments that make doing scientific research an exquisite privilege—that instant of realization that the ideas, the painstaking preparation, and the risk taking are going to pay off. It's like a private firework in your brain. It soon fades in the reality of still more preparation, fundraising, and the day-to-day routine. But the excitement of discovering something no one has ever seen before, succeeding at doing something that, as far as you know, has never been done, is a unique gift to cherish.

In the spring of 1996, we made our first attempts at using the ultrasound machine at sea in Cape Cod Bay. The weather was brutally cold and the water bumpy. The whales were doing long bottom-feeding dives, surfacing to take a few breaths before going down again. This behavior is typical of North Atlantic right whales in Cape Cod Bay in the early part of the year. But more proximate challenges also played a role in our self-doubt. The boat we were using had an upper steering station with a single waist-high aluminum rail surrounding it. When a 3-foot (1 m) swell sent Carolyn and me sliding on the icy upper deck, we resolved to develop the method in a warmer season. Still, we were able to acquire ultrasound data on live whales that spring, and were encouraged to carry on with our quest.

The nearest location where right whales were known to be accessible in summer and fall was Nova Scotia. Right whale sightings south of Nova Scotia, in Roseway Basin and east and north along the Scotian Shelf, were documented in the 1960s. There were also right whale sightings between Bar Harbor, Maine, and Yarmouth, Nova Scotia, in the late 1960s. Then, in the 1970s, they were seen in the Bay of Fundy, and have been studied there since 1980 by Scott Kraus, who developed and led the New England Aquarium's North Atlantic right whale program until 2019, along with Amy Knowlton and many, many others. They found the area to be a major source of sightings of North Atlantic right whales. These sightings added to those by Bill Schevill in 1955 in Cape Cod waters. So that fall, Carolyn and I stayed at the aquarium's research station in Lubec, Maine, locally known as the Whale House. At that time, North Atlantic right whales were commonly found feeding in the Bay of Fundy, in water 600 feet (180 m) deep, east of Grand Manan Island and west of Nova Scotia.

We successfully acquired blubber thickness echoes from two more North Atlantic right whales, but realized that the hand-held pole on which the ultrasound probe was mounted was too short. This was one of the many times that I learned the lesson that you can develop technology with or without engineers. The latter approach rarely works well, so bite the bullet, raise the money, and pay engineers to do it right. I worked with an engineer at WHOI who designed and fabricated a new articulated pole, attached to the boat, that worked much better than the shorter hand-held pole.

March 1997 saw us back in Cape Cod Bay for the first trials of the new articulated pole. It took a fair amount of tweaking, but we ended up with a system that appeared to have promise, although despite some attempts, we failed to acquire any blubber thickness data. The major issue at that stage was that the pole had a series of stiffening guys that made it hard to handle, and also added weight to the pole. In reality, it was now overengineered. Thus, the other part of the lesson: engineers can be too much of a good thing. My brother-in-law, who had a lot of experience with rigging carbon fiber sailboat masts, told me to ditch all the rigging that was encumbering the pole and use it unsupported. This advice made a huge difference in terms of operability, and the pole was still remarkably stiff.

In August 1997, we took the boat back to Lubec and finally did well, acquiring blubber thickness data from 17 different animals in a single day. In addition to having the right amount of engineering, we had also found the right pole operator. He was a colleague who had grown up playing baseball and bullraking clams. His ability to place the ultrasound probe on the back of a whale accurately, whenever I was able to maneuver the boat to a position where he could do so, was uncanny. The third and most important ingredient in our success was having relatively flat water with lots of approachable whales. Given that they were diving down as much as 600 feet, they were out of breath when they came back up and needed to spend a good deal of time at the surface recovering, which allowed us to gently follow them and place the probe on their backs as they surfaced. As time went by, we slowly learned how to approach the animals in a way that did not seem to disturb them. The actual touch of the ultrasound probe on the back did not seem to be noticeable if there were waves or ripples on the water, but if the water surface was mirror calm, the animals seemed to feel it.

We were doing well in terms of getting to the right place to acquire data, but the ultrasound unit we were using was not generating particularly crisp data. That winter, I worked with an acoustic engineer at WHOI to upgrade the ultrasound. Thus, when we returned to Lubec in 1998, we had a functional system and were able to start collecting a consistent data series. Each time we used the ultrasound on a whale, we also took photographs of its head so that we could compare them with the ID catalog at the aquarium. Knowing the identity, and therefore the history, of an individual made the data we acquired far more useful. Specifically, a year or more after we acquired the blubber thickness data, we could go back to the catalog to see which of the animals we had sampled were soon to get pregnant, or not, or were pregnant at the time they were sampled. In this way, Carolyn was able to show that North Atlantic right whales did not get pregnant unless they were fat enough to wean a calf successfully. It made us start to wonder about the impact of stressors that reduced their body condition. Obviously, lack of food would be a cru-

Approaching a North Atlantic right whale to measure blubber thickness using ultrasound in the Bay of Fundy, 1999. *Top*: The cantilevered pole, with ultrasound probe at the far left. *Bottom*: The probe gently touches the whale's back to acquire echoes of the blubber-muscle interface. Photo: Woods Hole Oceanographic Institution.

cial one, but unplanned energy costs, such as the added drag from persistent fishing-gear entanglement, could also be sufficient to tip the balance such that the whales lost too much energy to get pregnant.

In parallel with our blubber analysis in the Bay of Fundy, a group from La Jolla, California, working for the National Marine Fisheries Service, was using a plane to measure the same North Atlantic right whales. Carolyn used their photographs to measure the length of each whale and its width at multiple points. With the aerial data calibrated by our measurements of blubber thicknesses, we could use aerial images acquired each year to monitor the health of individual whales, and in so doing, get a sense of the changing health of the species through time. This approach has only improved with the advent of small drones.

While we were in Lubec in 1997 measuring blubber thickness in the middle of August, the New England Aquarium's research vessel *Nereid* was en route to survey for North Atlantic right whales in the Bay of Fundy. The crew heard a radio call asking if "the little green boat that chases whales is out today." The call came from a tuna boat, *Twilight Mistress*, reporting a floating dead whale. The location of the whale was just east of the shipping lanes leading to Saint John, New Brunswick, about 8 nautical miles from the Nova Scotia coast. The *Nereid* diverted to search for the dead whale, located the carcass, and confirmed it to be a North Atlantic right whale.

The next day, we caught a ride in the aerial survey plane from the grass airstrip in Lubec, cleared Customs on Grand Manan Island, New Brunswick, and continued on to Brier Island, Nova Scotia. The animal was towed into Flour Cove on Long Island, which is adjacent to Brier Island. My goal at the time was to measure the thickness of the dead whale's blubber coat to ground-truth the work that Carolyn was starting to do. I met Bill McLellan for the first time when he emerged from the fog, crawling out of the water in a wetsuit, over slippery cobbles, with a line in his teeth. At the other end of the line was the dead whale. It was then hauled up the beach by a heavy bulldozer. Bill had been trained in whale necropsy techniques at the Smithsonian Institution and by then was at the University of North Carolina Wilmington. Pierre-Yves Daoust, a veterinary pathologist from the Atlantic Veterinary College in Prince Edward Island, was also there. They both taught me much about how to take large whales apart. The necropsy identified a large bruise on the left flank, with evidence of the right lower jaw being fractured at some time before the animal died. It was matched to #2450 in the catalog, a female that had first been seen in 1994 at an unknown age.

The circumstances of this animal's demise were to become quite familiar: broken jaw, heavy bruising, and proximity to a shipping lane. The best interpretation was that the whale had been hit by a ship and succumbed to the resultant injuries some days later. This was the first of three North Atlantic right whale carcasses that I was to look at in Nova Scotia. Of the other two, one had been seen off Cape Sable, then drifted on the fierce Bay of Fundy tides until it was finally towed in to Digby, with a broken jaw and a massive crack in its skull. The Canadian Coast Guard towed another into Kelley's Cove, Nova Scotia, about 5 miles south of Yarmouth, a seasonal ferry port connected to the United States, at that time by a regular ferry from Portland, Maine, and a fast ferry from Bar Harbor, Maine. It had many broken bones in its back.

When news of a dead whale arrived at Woods Hole, we often hit the road in my old pickup truck, gathering crew as we sped north, hoping we had not left any critical gear behind. The wonderful thing about doing North Atlantic right whale necropsies in the Canadian Maritimes at that time was Jerry Conway. Jerry worked for the Canadian government, and part of his job was to gather data on North Atlantic right whales. He had an uncanny ability to quietly organize the necessary logistics for getting dead whales to the beach and hauled up above the high-tide mark, so that usually by the time we arrived, breathless, but keen to look at the animal before it decomposed any more than it already had, we were able to get straight to work. We would often arrive at dawn, and there would be Jerry with that smile of his, having got it all organized for us. We would then do what had to be done with what we had available to us.

. . .

As the years rolled on, our community of whale biologists and veterinarians evolved a system for managing dead whales on a beach so as to acquire useful diagnostic information that was relevant to managing the sources of the industrial trauma that we were documenting.[4] The stranding network on the US east coast includes a series of nongovernmental organizations that are authorized by NOAA to manage live and dead stranding events for specific coastal areas. For instance, I worked primarily on the south coast of Massachusetts, which included Cape Cod and the Elizabeth Islands. But those of us who had gained significant experience with examining dead right whales were invited to work on carcasses in other locations to get all we could out of them. This resulted in my gaining quite intimate knowledge of beaches from the Gulf of St. Lawrence to Florida, often working closely with Bill McLellan, Sue Barco from the Virginia Aquarium, and many others. Case records from the 1970s focused on anatomy, but as time went by, we increasingly focused on diagnosing cause of death.

As I had learned on Cape Cod, it is critical to get the animal above the high-tide mark so that we can look at it without the tide and surf damaging it further. We do this with one or more heavy machines on the beach if the whale can be buried there, or we tow it to a port, lift it out of the water, and truck it to a suitable facility, such as a landfill or, occasionally, a composting facility, if there are options to bury or compost it inland. So much of our logistical planning is dictated by the options for disposal of the carcass. Disposal at sea has been discussed over the decades, but we have never attempted that for large whales. A floating carcass creates an ongoing navigational hazard. It is also at risk of coming ashore again:

beached carcasses have refloated and have then been sighted far offshore before coming ashore again somewhere else.

Once we have decided that we have a viable disposal plan for a carcass, we then start to look at what resources we have in terms of people willing and able to examine it. The temptation is to grab some knives and start taking it apart, but there are a number of critical steps to follow first.

My first step is to gather all of the available data about the history of the case: when and where the carcass was first sighted, who did what to it, what we know about it. Photographs taken at that time are critical, in that they are our best chance of observing the case before further decomposition masks essential evidence of how the animal died. I need to know the origin of any rope on a carcass. Was it part of the cause of death, acquired during a lethal entanglement in fishing gear, or was it wrapped around the animal after death by wave action at sea, or was it added as a towline to bring the animal ashore? If there are rope marks on the carcass, were they inflicted during the animal's demise, or postmortem by a person working to retrieve it? Thus, I try to talk with anyone involved in a case.

Once the animal is beached, I do an unhurried, thoughtful visual examination of its surface. Are there any lesions: fresh wounds, healed scars, parasites, individually recognizable skin pigmentation, or other identifying marks, such as callosities on a North Atlantic right whale? Callosities will fall off during the decomposition process as the skin detaches. There can be very subtle patterns embedded in the skin surface. For instance, when a gill net rolls up during an entanglement, its fine monofilament mesh can make complex indentations on whale skin. The telltale pattern is many intersecting parallel lines going in two different directions, but usually not perpendicular to each other, as the bunching up of the mesh also involves a stretching of the net into a series of wound-up parallelograms, which emboss the skin with two sets of lines at a specific angle. These marks are often quite cryptic, but they become obvious once you see the patterns. To do this, I walk around the animal, in both directions, a few times, ideally at different times of the day, as changing light angle and amount of cloud cover can either hide or highlight the evidence. If there are multiple lesions evident, we will assign a number to each lesion. That number is recorded on data sheets, photographs, and samples for analysis. In that way, observations from all of the stages of analysis can be correlated once all the data have been assembled for review. Quite often we spend the better part of a day getting the carcass to the beach and hauled out of the water. We then spend the rest of the daylight documenting what we can see and photograph, before returning the following day at dawn to take the animal apart.

Dissecting a large whale carcass can take less than an hour on a whaling station or ship deck, with highly trained workers wielding razor-sharp flensing knives and multiple cables and winches. They remove the blubber as if they are peeling a banana, then pull out giant steaks of muscle along the vertebral column, moving aside the viscera, before using giant reciprocating steam saws to cut the skeleton into manageable chunks. Or dissection can be a challenging task, with a few biologists poking at the carcass with a limited supply of rapidly blunting knives and no heavy machinery. But even in the latter scenario, with an incomplete dissection, biologists, with thoughtful focus on critical areas, can often accumulate enough evidence to understand the cause of death. Happily, in the past decade, the number of whale pathologists in Canada has increased along with the growing number of dead right whales found there. We have also been fortunate in recent years to have the support of the federal governments of the United States and Canada, which have provided suitable heavy machinery to assist with the disassembly task. Excavators with articulated buckets attached by chain to blubber, muscle, and bone can enable the disassembly of a North Atlantic right whale in less than a day, assuming there are a few experienced people able to make the necessary cuts, and able to work with the heavy machinery. A critical skill set for a person leading a large whale necropsy is knowledge of the gross anatomy of the specimen—where to cut to maximize the efficiency of the tension available from the heavy machinery. Cutting, tearing, and pulling in the right planes of tissue cleavage is critical to get into and around the various structures of the carcass.

When we take large whales apart to find out why they died, we sometimes remain none the wiser. But quite often, we learn a lot, even from animals that are heavily decomposed. Important tissues such as blubber and bone are highly resistant to decomposition, and they retain critical information.

There are two primary classes of trauma caused by vessels: cuts made by rotating propellers, and impacts from a bow, a hull, or another relatively blunt structure, such as a rudder or stabilizer. Propellers will leave one or more curved incisions that, if multiple, form a series of parallel cuts. The cuts may penetrate only the skin and part of the blubber, in which case they are often not lethal. Or, if the propeller is large compared with the size of the whale, it can cut much deeper into internal organs such as the jaw, skull, chest, or abdomen. The latter cuts are often lethal. Propellers can also cut off part or all of the tail flukes. Cuts in the flukes can be lethal if they cross the midline of the animal, as they will sever the major blood supply to the tail, and blood in such large vessels cannot clot fast enough. Or the chest can be breached, leaving the animal unable to breathe. Blunt trauma often leaves obvious bruising in the blubber, which may overlie a fractured spine and severed spinal cord, leaving the animal unable to swim. It also can lead to massive fractures in the skull, involving the braincase or a jaw. Most of these whales die relatively quickly: one North Atlantic right whale that had an entire tail fluke cut off died within an hour, as it bled profusely.

One series of propeller cuts to a North Atlantic right whale fluke that was not immediately lethal was seen months later; the animal was in poor shape, and it was never seen again. Perhaps the saddest case was that of a North Atlantic right whale called Lucky. In January 2005, her carcass was spotted 30 miles east of Cumberland Island, Georgia, by a New England Aquarium survey plane. It washed ashore just north of Jacksonville, Florida, three days later. Lucky, then a pregnant female, had three healed propeller wounds. Those wounds had been present at the animal's first sighting as a calf in 1991. She got her name because observers felt she was lucky not to die from the wounds at that time. At her death, there was evidence of inflammation and infection associated with the wounds. The cause of death was unclear, although it seems quite likely that her advancing pregnancy had stretched the wounds, which in turn led to an overwhelming infection.

There are many known cases of whales surviving propeller trauma. We know far less about animals surviving blunt trauma, because although evidence of internal bruising and fractures can be observed during a necropsy, there is rarely any external evidence for such a collision in a live animal. Most cetaceans have dark skin, which makes surface bruises cryptic. However, with Lucky as an exception to the rule, if either type of vessel trauma is going to kill the animal, it usually does so quickly.

. . .

Like trauma from vessel collisions, entanglement in fishing gear has a range of outcomes. Whales can survive, or die quickly or slowly. If the gear that entangles the animal is anchored to the bottom well enough, or is heavy and strong enough, and the water deep enough, the whale will not be able to reach the surface to breathe, so it will run out of oxygen and build up too much carbon dioxide. Death will not be instantaneous, because these diving animals have excellent breath-holding capacity. Thus, the time it takes for them to asphyxiate will probably be substantially longer than the length of a normal dive—perhaps a half hour for North Atlantic right whales, as they often dive for about twelve minutes, and two to three hours or more for sperm or beaked whales, given that they routinely dive for an hour or more.

If the whale is strong enough to break out of the entangling gear or swim it up to the surface, then it will be able to breathe. It may then have the energy and oxygen to wriggle out of the gear and carry on with its life, bearing "only" wounds from the entanglement, which will hopefully heal. The vast majority of North Atlantic right whales have scars from rope and net entanglements.

However, quite often, the animal can swim to the surface, but is left dragging rope, nets, and traps as it continues to move, depending on the nature of the gear it encountered and the degree to which it is able to disentangle itself. If it is dragging significant amounts of material, it risks a worsening condition. The few times whales have been observed in the early stages of an entanglement, they exhibit extremely agitated behavior, rolling, thrashing with their flukes, traveling very fast, and generally trying to shake off the gear. This behavior may help, but it may also hinder. A whale's body

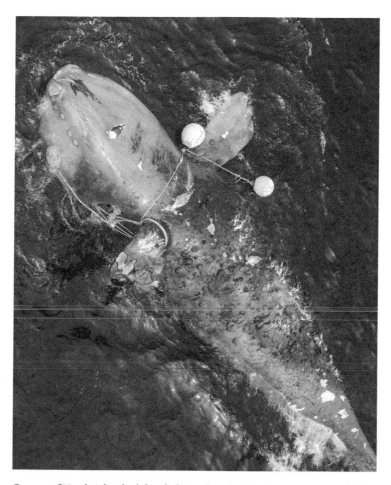

Carcass of North Atlantic right whale Starboard (#3603), June 22, 2017, with rope cleated between the right flipper and the inside of the mouth. It probably drowned in Canadian snow crab gear. Photo: NOAA/NEFSC/Peter Duley. Permit: SARA DFO-MAR2016-2.

has a number of places for gear to get caught: the upper and lower jaw, the baleen plates, the two flippers, and the tail flukes. The more it rolls, the more turns of rope it gets around these various body parts, and the harder it swims, the tighter the wraps become. Once the entanglement involves more than one body part, with one acting as a cleat, the wraps of rope tighten and constrict as the animal continues to swim. At times, random knots also form as the animal moves. The flukes and peduncle pump up and down, the body flexes and extends, the flippers and flukes move to control the direction of the animal, the head moves up and down and side to side, and the mouth opens and closes. All of these movements cyclically tighten and loosen the rope and its wraps around the body part. If the animal does not work free, there is a major risk that the rope constrictions will slowly get tighter and tighter, pressing down into blubber, muscle, and bone, killing tissues, and eliciting a scarring response from the animal that tries to wall off the entanglement, treating it as a foreign body, as any mammal would.

One case we examined in Virginia, North Atlantic right whale #2301, had been first sighted entangled on September 6, 2004, on Brown's Bank, southwest of Nova Scotia. The crew reported her to be in poor health, with pale, patchy skin and whale lice spread over her body, especially in wounds around the blowhole. These lice are usually confined to the head callosities and body crevices. The left flipper was white, and the animal appeared thin. They reported a single sun-bleached green rope exiting from two points on the right side of the mouth and running over the blowhole, then down to the left flipper. It was assumed that the rope wrapped around the flipper was causing the white color.

A disentanglement team was assembled, and three weeks later they were able to head out to the area to look for the whale. She was the first whale they found that day. They got into cutting range three times, but never loosened the wraps, although they felt that they had cut at least one line. The team confirmed that the left flipper was fully white, an indication of severe constriction at its insertion by the entangling ropes.

They further noted that the entangling lines causing this condition were not visible, probably indicating that they were wrapped around the flipper and deeply embedded. They also found that the loss of pigmentation in the flipper extended onto the body of the whale. A previously unreported length of green line was seen trailing beneath the whale as far back as the flukes. The next time the whale was seen alive was on December 8, 2004, off Nags Head, North Carolina. The entangling ropes were still evident at that last sighting. The whale was then found dead on Wreck Island, Virginia, on March 3, 2005—nearly seven months after the initial sighting.

That day, I was attending an annual meeting of the Northeast Regional Marine Mammal Stranding Network in Virginia Beach, Virginia. We stayed in one of the cookie-cutter beachfront hotels to which vacationers flock, which form glass and concrete high-rise cliffs from New Jersey to Florida. Heading into breakfast, I was stopped by our host, Sue Barco, who at that time responded to marine mammal strandings in Virginia. She told me about the whale on the beach, and asked me to go with a crew from her group to see what we could learn. This mission involved a 40-mile drive north, trailing a boat, over and through the Chesapeake Bay Bridge-Tunnel to the community of Oyster, a very different place from Virginia Beach. Oyster was the heart of Virginia's oyster business and featured long, elegant dead-rise work boats, low-lying islands, birds galore, and drying creeks and sandbars. We launched the boat and headed out through Sand Shoal Channel to the southeast corner of Wreck Island, 8 miles east of Oyster, a shifting, low barrier island bordering a massive creek system to its west. Access to the site was difficult, even in the small boat we were using, given the 2- to 3-mile-wide mess of breakers and shoals east of the island and the shallow tidal flats to its west. There were four of us, with knives, a few shovels, a tape measure, and a camera. It was cold and wet. I had not yet fully recovered from about six months of poor health. It was all very depressing, but we did what we had to do.

The animal was half buried in the sand on the eastern, exposed side of the island. It must have been washed over by waves and swells after it came ashore. It could have been there some time, undetected, as it was in a deserted, lonely place. To get the carcass anywhere to do a proper necropsy was out of the question. It would need an excavator brought in by barge to dig it out. Larger boat or barge access was out of the question. So, we got to work, photographing what was visible, and then started digging. Fortunately, the animal was tilted so that the left side was less buried than the right. The sand, mostly ground-up oyster shell, was like concrete that had almost cured. It was the hardest digging I had ever had to do. We really wanted to look at the right flipper as well as the left, to see if it, too, was entangled, but it was under the animal, and so deep that we would never have reached it in days of hand digging in that soggy semi-concrete.

As we started to expose the left flipper, the true magnitude of the entanglement became horrifyingly clear. It showed evidence of major constrictive forces from multiple wraps of rope. There was a V-shaped wedge where the skin and fibrous soft tissue that normally overlie the flipper bones were gone. They were replaced with massive proliferation of crumbly new bone forming on the upper bones of the flipper. There was also a major furrow running across the left blowhole. But the rope that had been documented crossing the blowhole and running down to the left flipper, after exiting the mouth, when the animal was alive was no longer present. Once we were able to examine the inside face of the left baleen, we did find a mass of tangled, knotted rope woven between the plates of baleen. Some of the rope ends were frayed, suggesting they had broken under tension, whereas others had cleanly cut ends, most likely cut with a knife. The frayed ropes probably broke when the animal was first entangled. The knife cuts could have been part of the September 22, 2004, disentanglement attempt described above, but the rope that was seen exiting the right side of the mouth and then crossing over the blowhole must have been lost, or perhaps removed, after the whale was seen in North Carolina, either when floating dead at sea or once it had beached, before we examined it. Perhaps the rope detached during the beaching process. However, the salient aspects of the case were clear. The animal was alive, entangled, for three to six months while it swam a minimum of 800 nautical miles, from New Brunswick to North Carolina and back to Virginia, with a complex mess of knotted rope in its mouth and baleen, over its blowhole, and wrapped around its left flipper. As the animal swam and flexed and extended its body, it pulled back and forth on the taut line, which carved a furrow in the left blowhole, dissected the flesh from the upper part of the left flipper, and caused the massive proliferation of new bone on the surface of the exposed flipper bones. This animal must have undergone months of chronic pain. In contrast to most whales fatally struck by vessels, entangled ones can linger for months or years. These longer time periods put the seconds to minutes it took for harpoons to kill fin whales in Iceland in a new perspective.[5]

By this time, I had a pretty good sense of the devastating effects on animal welfare that these chronic large whale entanglements represented. To take an animal apart, and slowly piece together the story of how it had died, was not easy. We had to think through each step of the events that animal had gone through. I resisted trying to put my mind

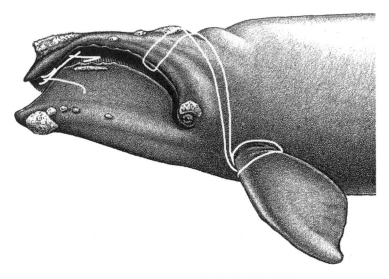

Sketch of the rope entanglement on North Atlantic right whale #2301.
Drawing: Scott Landry, Center for Coastal Studies.

into the animal's head, but to really understand what these animals experience, such an exercise is a necessary evil. As I got my head around the tangled mess that had killed whale #2301, and the mechanics of it all, the veterinary side of my brain formed an angry, confused resolution to do better by these whales.

The veterinarian in me began to transition. Veterinarians are trained to manage a case by first acquiring salient aspects of the case's history, then examining the clinical signs and lesions present to establish a diagnosis. This diagnosis then leads to a treatment plan, and ultimately, to prevention of recurrence of the problem at the individual and herd scale. History/examination, diagnosis, treatment, and prevention. I had begun to understand what led up to this kind of event. I was taking note of the history of each case, and through cases such as that of whale #2301, the nature of the problem had become evident. I was finding the same value in my veterinary training that Tom Albert had in Barrow many years before. As I began working alongside disentanglement crews, it was a natural progression to consider the possibility of treatment, and the potential for preventing the problem.

6

Treating Whales

In the late 1990s, I was under the naive impression that the research on whale trauma caused by vessel strikes and rope entanglement, done by colleagues such as Scott Kraus, Amy Knowlton, and Bill McLellan, would lead to changes to the shipping and fishing industries in the United States and Canada, and that those changes would significantly reduce large whale trauma. After all, in the United States at least, the Endangered Species Act and the Marine Mammal Protection Act were at least twenty-five years old. I assumed that whale biologists such as myself would still have a stopgap role to play in continuing to document causes of death, and that my colleagues would continue to disentangle whales as they could, but that prevention of these traumatic events was just around the corner. How wrong I was. In August 2020— more than two decades later, with those same laws close to fifty years old—a judge ruled on a lawsuit seeking to close an area south of Nantucket to lobster fishing until the National Marine Fisheries Service followed the rules and met its obligation to enable the recovery of the North Atlantic right whale species, albeit five decades after federal laws first required it. All the judge did was kick the ball down the road for another nine months. I tweeted:

Can North Atlantic right whales hold their breath that long?
1996 Take Reduction Plan allowed NMFS 6 months to fix
their problems. Most recent ruling, TWENTY-FOUR years
later, gave NMFS another 9 months.[1]

The irony does not escape me that I have been examining,
and thinking about, dead right whales for those same twenty-
four years.

In chapter 8, I will expand on what needs to change, but it
is patently obvious that relying on the current laws and regulations of the United States and Canada has yet to ensure
conservation of the North Atlantic right whale species or prevent severe, prolonged suffering for individual animals. How
can this change? The politicians who manipulate the management of such problems by government employees need to
understand that a majority of their electorate really does care
about these issues. Voters must urge politicians to genuinely
enable suitable and effective conservation measures while
still supporting the livelihoods of those employed in affected
industries such as fishing and shipping. This is true in the
case of the right whale, and it is true for many other situations
in which there is conflict between short-term individual and
corporate gain and the sustainability of diverse ecosystems
across the globe in the much longer term.

Meanwhile, until we can get to such a place, treating the
effects of large whale trauma continues to be an essential part
of whale conservation and welfare. Indeed, genuine prevention of the problem is still in its infancy. With that depressing
background, I can now tell you about the whale that was for
me the poster child of the right whale entanglement problem
in 1999.

. . .

On May 10, 1999, an NMFS survey plane had observed right whale #2030, a female, in the Great South Channel, east of Nantucket, with multiple wraps of net and its associated lines around her body. She was not sighted again until September 3, 1999, in the Bay of Fundy. By then three wraps of line remained, one of which cut deeply into the blubber. Crews disentangle most whales by attaching large buoys to line trailing behind the animal so that the added drag will tire the whale, allowing them to approach it and make the necessary cuts to remove the gear. Sometimes trailing line from the whale is bent over the tube of an inflatable boat to result in more drag, and/or to help the rescuers pull themselves closer to the tail of the whale. But this whale lacked sufficient trailing line to do that. However, rescuers were able to attach a buoy with a satellite tracking tag to the stub of line that was present. On September 4, some of the line was cut by Stormy Mayo and David Mattila from the Center for Coastal Studies.[2] The animal was evasive and difficult to approach, and the conditions were foggy. For the next few days, bad weather precluded further efforts. Stormy and David returned to Provincetown, where they developed a prototype tail harness to lasso the whale by the flukes.

The following week, I was back in Woods Hole when my office phone rang. It was Teri Rowles, a veterinarian with the NMFS Marine Mammal Health and Stranding Response Program. Whale #2030 was still in rough shape, given that the gear had cut into the blubber across her back. Could I adapt the pole system we had developed for measuring blubber thickness with ultrasound so that it could be used to deliver antibiotics to #2030, if they were able to disentangle her? Teri has a habit of asking piercing questions for which there are

North Atlantic right whale #2030, May 10, 1999, observed in the Great South Channel east of Nantucket. The net and associated ropes are wrapped over the back between the flippers but are not cutting in at this point. Photo: NOAA NMFS North East Fisheries Science Center.

rarely simple answers, and which lead to physical and creative places where I would never have thought of going myself. Her focus is invariably on the best interests of the animal or animals in question. Her question led me down the hall to my neighbor, Craig Taylor, a deep-sea biologist with experience in remote sampling from WHOI's manned submersible, *Alvin*. I was in search of a remotely triggered syringe that we could attach to the pole. I felt rather ridiculous verbalizing that I needed help to give a large whale an injection at sea. Craig smiled and pulled a 1 L syringe, with a spring-loaded plunger, off a high dusty shelf. I walked the syringe across the drawbridge outside our lab building to a colleague, and engineer, Terry Hammar. I needed a long needle attached to the syringe. We had figured that 12 inches would pass through the 8 or so inches of North Atlantic right whale blubber into the muscle, so that we could deliver an intramuscular injection.

It's hard to describe the conflicting senses of excitement, trepidation, and intense focus that arise in situations where one is trying to figure out how to do something that seemingly no one has succeeded in doing before. What will go wrong? Where are the safety issues? What will other people think? What haven't we thought of? Little did I know that this effort was the first step in a chain of consequence that still has not found its end. We had begun our first attempt to treat a whale at sea to aid disentanglement efforts.

Terry had duly welded up a suitable needle that attached to the syringe. A line was attached to its spring-loaded release so it could be manually triggered by the pole operator. Back on the road to Lubec, towing the boat we had been using for the blubber ultrasound project behind my pickup truck, we were as ready as we could be. Late that night, we met in Lubec with the disentanglement team that had driven up from Provincetown. David Mattila explained the plan for use of the tail harness, and I described the proposed procedure and equipment for injecting antibiotics into #2030 if she was approachable. We had gained approval from both US and Canadian officials, despite both methods being untested, given the critical nature of #2030's entanglement.

The next day, the New England Aquarium vessel *Nereid* departed Lubec at dawn with rescue team and gear onboard. A small inflatable and our boat soon followed. Two hours later, we all met up with a sailboat that had been radio-tracking the whale. An inflatable was set up for deploying the tail harness. We watched the various boats interact with the whale while we made sure we were ready to deliver antibiotics if and when the animal was disentangled. Most of the day was spent attempting to attach the tail harness to the whale. After multiple unsuccessful attempts in less than ideal conditions, the rescue was called off for the day. Despite her outward appearance of poor health, the active behavior of the whale belied her assumed deteriorated condition.

The following day, Stormy and David tried to slow the whale down with sea anchors. The animal continued to dive and evade attempts to help her. Tropical Storm Floyd's approach made it impossible to continue, and we had to return to Woods Hole. The team continued to track #2030 after she left the Bay of Fundy. She was south of Nantucket by September 19. By September 27, the satellite tracking buoy no longer seemed to be attached.

That fall, I prepared three presentations for the 1999 meeting of the North Atlantic Right Whale Consortium.[3] The consortium has met annually since 1986 to review the status of the species, communicate new findings, and discuss conservation efforts. The meeting that year was at the New England Aquarium. Carolyn Miller and I were to present our findings on blubber thickness in North Atlantic and Southern right whales, along with our techniques of using an ultrasound probe on a cantilevered pole to measure their blubber thickness and using aerial images to measure their length and width. I had also prepared a talk on pursuing Teri Rowles's concept for drug delivery. My theme was that recent life-threatening entanglements and potential disease reports from North Atlantic right whales had prompted us to consider options for medical intervention in free-swimming large whales. Because of the species' endangered status, it seemed worthwhile going to extreme efforts to save an animal's life. But these whales were not relaxed patients, and were an order of magnitude larger than the largest terrestrial animals subject to veterinary attention, and no one had developed either the technology to deliver drugs to or the techniques to handle

an animal of their size. In the last six months, we had identified potential applications for sedatives, antibiotics, steroids,
and local anesthesia. We proposed that a workshop be held
to consider these options, to develop strategies for rapid and
remote drug delivery systems, to evaluate appropriate drugs
and concentrations (including consideration of the scaling
issues), and to make recommendations on further development.

The meeting was to start on October 21, but on October
20, the US Coast Guard received a call from a sailing vessel
reporting a floating dead whale "that was 25'-30' in length,
gray and black in color and tied up in nets." It was eventually
identified as a North Atlantic right whale: #2030. This put me
in a real quandary. I wanted to examine #2030 to understand
how the gear had killed her, but I also needed to make the
presentations at the meeting. Some very experienced people
were willing to go, so I decided to stay in Boston.

The US Coast Guard towed the whale to its Cape May,
New Jersey, base, and a crew traveled down to examine the
animal. As the whale had been found with its back out of the
water on October 20, they concluded that the animal had not
been dead long; once whales start to decompose, their intestines are the first organs to fill with gas, making the carcass
float first on its side, and then belly up. We later learned that
on October 16, 1999, the crew of a small boat had reported a
large whale (40-60 feet, or 12-18 m), entangled in fishing net,
2-3 miles off the coast on the Ship Bottom shoal, and had told
the Coast Guard that the whale was "acting aggressively."
Thus, the animal had to have died between October 16 and
October 20.

The lines wrapping around both flippers had cut down
into the bone at the leading edges.[4] There were knotted
tangles at each flipper leading to a single embedded line over

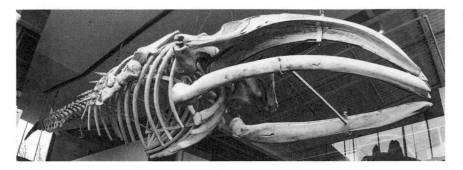

The skeleton of North Atlantic right whale #2030, carefully cleaned and displayed at the Paleontological Research Institute in Ithaca, New York. Photo: Author.

the back, and two tight lines around the chest. Gear trailed from the left flipper. By this time, the line over the back was embedded in the back muscles, which no longer had any covering blubber for a distance of 4.6 feet (1.4 m) along the back. As the cut ran laterally, the gap tapered until closing at the flippers. It was as if the whale had been wearing a constricting, tissue-melting shawl of gill net and rope, spread out over the back and tucked in tight around both flippers. There was evidence of shark bites at the wound edge. As the animal was hauled up the beach, the left shoulder blade partially fell away from the carcass. The lines were under extreme tension and snapped back when cut. The bone of the inside face of the right shoulder blade had grown around the rope. The remaining blubber was very thin, showing that the animal was emaciated. Internally, there was no evidence of traumatic skeletal damage or bruising in soft tissue. The diagnosis was massive traumatic injury induced by entanglement in fishing gear with resultant starvation.

I summarized that case two years later, after going through all of the available reports and data. To watch the animal swimming in the Bay of Fundy with the line stretched between her armpits, with line and net cutting down into the blubber on her back, was deeply disturbing. The realization that subsequently, while she was still alive, the line had continued to strip the blubber off her back made it hard to comprehend the suffering the animal must have sustained. She had been entangled for a minimum of six months. She became a ghastly icon that surfaced repeatedly in my mind and drove me, and many others, I suspect, to work relentlessly to avoid the recurrence of such a case. It was hard not to be deeply sad. While the veterinarian in me was seething with indignation that our consumer demand for seafood was driving this problem, I had no angst toward the fishing industry. It was simply doing what I, as one of many, many seafood consumers, demanded. There had to be a better way.

Meanwhile, back in Woods Hole, Teri Rowles's idea that we could consider applying veterinary care to large whales suffering at sea had taken hold. Woods Hole Oceanographic Institution had a fund that supported interdisciplinary research by scientists and engineers who wished to collaborate and innovate. Terry Hammar and I submitted a proposal to develop a system to deliver drugs to live whales at sea. Our proposal included the convening of a workshop to seek input, advice, and ideas from veterinarians, whale biologists, and state and federal managers as to how best to do so. The proposal was reviewed positively, as the effort was considered timely in view of the extremely endangered status of the North Atlantic right whale.

In fall 1999, we started using the grant funds to develop the at-sea therapy concept. Getting a grant proposal funded always brings a complex rush of simultaneous emotions: relief that part of your job description has been satisfied, albeit in a relatively fleeting kind of way; excitement to be able to set out to do what you proposed, and trepidation over whether the premise, plan, and reality will yield a successful outcome. In this case, all of these emotions were intensified by the audacity of what we were proposing. Some years earlier, when I was gestating the idea of measuring blubber thickness using ultrasound at sea, I had run into Peter Tyack outside his office in the basement of WHOI's Redfield Building. He had moved there in 1982, working as a postdoc and then as a scientist. When I told him about my ultrasound idea, I sensed some hesitation regarding my sanity. Indeed, although he was outwardly encouraging at the time, he did admit years later that he thought I was out of my mind. Thus, when our proposal was funded, I was once again questioning my own judgment, worrying that I wasn't going to cut it, and wondering what on earth I thought I was doing. But when I sat down with Terry Hammar to talk through the operational and design constraints, he quietly helped me build confidence that what we proposed was doable. The third partner in the project was Andy Stamper, a veterinarian at the New England Aquarium. Andy was more of a clinician than I would ever claim to be, and he lent gravitas in terms of our credibility in that direction.

Complicating the execution of the project was that I had a plan afoot to take a year away from lab life with my family on a sailboat from June 2000 to August 2001. This plan was another huge test of my self-confidence, self-respect, and perspective on myself as a researcher at WHOI. Hannah and I had talked for years about how, when our four sons were in the eight- to twelve-year age range, we would like to take them on a sailboat for a year or so. Our hope, which in hindsight we achieved, was to build a sense of interdependence and mutual respect among the four of them and ourselves, in addition to showing them places and people that they would normally never encounter in Massachusetts. But again, my doubt as to the sanity of the plan was enormous. For me, the person who made it all right was Bill Watkins. One afternoon, I had been browsing some journals in his laboratory across the street from my office. I was looking for historical North Atlantic right whale sighting data for the Northeast Atlantic as part of my route planning for our voyage. Bill asked me what I was working on, and I sheepishly told him my growing plan. I mentioned my misgivings in terms of what an extended absence would do to my career. He stood up, but with his characteristic stoop, looked over his glasses, as he was wont to do, and asked me whatever possessed me to even consider not going. If Hannah and I could make it happen, of course we should go. Coming from Bill, who had a reputation as a workaholic with a single-minded focus on his science, that advice was a major turning point for me. So now I needed to figure out how to get the at-sea treatment project underway, and how to keep it going while largely out of touch and totally out of the office. My conversations with Terry and Andy were reassuring, as we figured we could get the bulk of the work done before I left, and Andy and Terry could do the fieldwork as opportunities arose. Our plan was made in close coordination with Teri Rowles, under whose NOAA Fisheries permit we would be doing the fieldwork.

One of the major tasks of the project was arranging the workshop at WHOI.[5] Twelve veterinarians, eleven biologists, a federal manager, and an engineer attended the workshop and helped us to refine our plan. There was consensus that available sedative drugs were potentially viable candidates for our project, whereas the antibiotic options at that time would not be effective as single-dose treatments. Chances of relocating animals at suitable dosing intervals were slim. (Longer-lasting antibiotics now allow single dosing as an option.) The justification for the use of antibiotics was also less clear than the case for using sedatives to make the animals more approachable for disentanglement and to enhance safety for both the humans and the whale. However, there was a stern warning in the literature, in a paper authored by Bill Schevill and colleagues, suggesting that what we were proposing could be, and had been, "disastrous." Admittedly, he was referring to attempts to anesthetize whales and dolphins, whereas we were planning light sedation, but Bill's words still gave me substantial pause: "In summary the use of presently known drugs to immobilize an aquatic mammal in the water is almost certain to kill the animal by suffocation."[6]

We initially focused on the drug Midazolam, widely used in humans to reduce anxiety and induce some drowsiness. We were very cognizant of the need to maintain a functional swimming, breathing, and diving whale patient.

Terry then designed a drug delivery system that was to be attached to the end of a long pole. We modeled the syringe on the principle used for ballistic terrestrial wildlife sedation darts. The major difference from the terrestrial version was volume: our system would have a 60 ml capacity, compared with the regular maximum of 6 ml. WHOI's engineering shop then fabricated the system.[7]

Shortly after the workshop, after I had got our sailboat to Ireland with some friends, Hannah, our four children, and I set forth to look for right whales in places they used to be. We visited old whaling grounds in Ireland, Scotland, the Azores, and Cabo Verde, and then sailed across to the eastern Caribbean before heading home via Bermuda, Newfoundland, and Labrador. We found no right whales, but we did build a very strong family bond, born out of prolonged, inescapable proximity, where issues could not be shelved. The motivation to agree to agree at sea is much greater than on land, where evasion and denial of issues can be easily achieved with a bit of distance. The real depth of the experience, in each of the Atlantic islands that we visited, was learning about the people and the place we were visiting. What had been the forces shaping the culture, how had the people prospered, or been abused historically, and what was going well or not so well for them currently? Perhaps the most meaningful experience for us was the six weeks we spent in the islands of Cabo Verde, south of the Canary Islands and west of Senegal. Living near New Bedford as we do, we all have many friends and colleagues whose families came from those islands. It was a common practice of nineteenth-century whaling ships from New Bedford to recruit skilled Cape Verdean seamen as harpooners and other crewmen. Consequently, there is a substantial, thriving Cape Verdean community in southeastern Massachusetts today. The islands are arid, hot, and starkly beautiful. Much as in the Arctic, the extreme environment has ensured that a tough, intelligent people survived.

While I was away with the family, Terry and Andy tested the whale sedation device successfully on samples of skin and blubber that had been collected from a dead right whale. Schevill had urged initial laboratory studies on live animals.[8] It is hard to get large whales into the lab, however, so the team first tested our drug choice on a beached whale that needed euthanasia as our first clinical step. In April 2000, a young, emaciated gray whale beached in California, and colleagues there administered Midazolam prior to euthanasia.[9] The effects of Midazolam were indeed to sedate the animal. But the only way to find out what the drug would do to a free-swimming whale was to inject a whale at sea. We waited until there was an entangled North Atlantic right whale that was very resistant to standard physical disentanglement approaches and almost certain to die if not disentangled.

Named for his refusal to give up, Churchill (#1102), an adult male North Atlantic right whale, was first seen by researchers from the University of Rhode Island in 1980 in Roseway Basin, south of Nova Scotia. The New England Aquarium saw him without any entanglement in August 1998 in the Bay of Fundy. On June 8, 2001—a bit more than a year after the workshop in Woods Hole—he was sighted from a NOAA survey plane east of Cape Cod. He was entangled and reported as having gray, mottled skin, with heavy green rope exiting the mouth and trailing 5 feet (1.5 m) behind the flukes. The plane relocated the whale the next day, and the disentanglement team from Provincetown arrived by boat. They documented the wounds, the rope embedded in, and wrapped around, the upper jaw,[10] and the parasites widely spread around the body. The outlook was grave. They attached a satellite tracking buoy to the rope to allow the animal to be found again after they had formulated a plan. After a lot of consultation, their assessment was that the best chance was a combination of physical and chemical restraint. Many preparations were made to maximize their chance of success. Essentially, all the efforts that had been made for #2030 were being redoubled. There was an overarching sense of frustration that these entanglement problems were persisting, and that while disentanglement efforts would potentially help individual animals, they were not going to resolve the problem facing the species. The practical challenges in this particular case included the relatively small amount of rope trailing behind the whale, which made it hard to manage him; his continued movement; his being underwater 80 percent of the time; and his distance from shore.

Aerial photographs on June 19 showed that the drag from the satellite tracking buoy had pulled the rope around the

Delivery of sedatives to Churchill (#1102) using a large syringe hinged on the end of a cantilevered pole. Photo: Center for Coastal Studies. Permit: NMFS #932-1498.

upper jaw, raising the hope that it might continue to a point where it came off the animal, but there appeared to be a tangle that would prevent that outcome.

The first attempt to sedate Churchill was on June 26, 2001. The plan was to start with a very low dose, observe what happened, and then incrementally increase the dose until the desired effect of making the whale more approachable was observed. The whale showed no behavioral reaction to the actual injection, but likewise, the low dose had little effect on him.[11] This was good news, in that we now had a safe baseline. The next step was to increase the dose. On July 14, the whale was 86 nautical miles east of Provincetown. This time, a higher dose of two drugs was delivered, twice over two hours, but again with no effect. The overall condition of the animal was worsening. On July 17, he crossed over into Canadian waters. Churchill then traveled south of Nova Scotia. By the end of the month, he moved into the Cabot Strait and then into the Gulf of St. Lawrence. Ironically, my family and I crossed paths with him at that point, as we were headed home. We had spent July in Newfoundland and Labrador, and we arrived home on August 5. Thus, we and Churchill had been in the Cabot Strait at the same time.

By August 2, Churchill was east of the Magdalene Islands, and by mid-August over the Canadian portion of Georges Bank and the associated canyons to its southwest. Then, on August 29, the animal moved close enough to Cape Cod that researchers could again attempt to disentangle him. By this time, I was back at work and participated. We delivered the drugs Midazolam and Meperidine, but disentanglement attempts were not successful.[12] It was a strange day for me. Getting back into the swim of North Atlantic right whale conservation work was a challenge, as I had been focused on my family for the previous year. My colleagues had put an enormous amount of effort into trying to save the animal over the previous three months.

Late that evening we returned to Provincetown, hungry, tired, and depressed. To be looking for supper amid the throbbing, swirling madhouse of late-summer downtown nightlife, after having spent the day at sea with an emaciated, entangled, suffering, dying North Atlantic right whale, was truly otherworldly. But in another way, walking off the municipal pier into the town was supremely poignant. The de facto world center for large whale disentanglement is in Provincetown, at its Center for Coastal Studies. And the town had spent the previous two decades living and dying through the AIDS crisis. Deep down, Provincetown shared our understanding of the enormity of the burden of terminal, chronic suffering.

Churchill's tag track ended abruptly on September 16, 2001, and we assume that the animal and tag sank 400 miles east of New Jersey, in 15,000 feet (4,600 m) of water. At the point

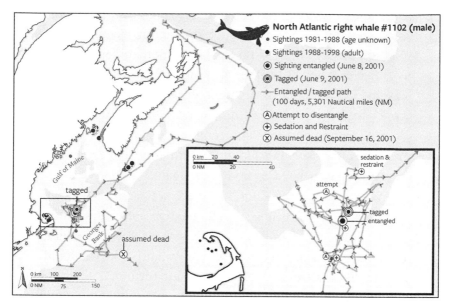

Telemetry tag track for North Atlantic right whale Churchill (#1102) while entangled. Data: Center for Coastal Studies. Permit: NMFS #932-1498. Plot: © Woods Hole Oceanographic Institution, Natalie Renier, WHOI Creative.

of the last satellite fix, the tagged whale had logged 4,929 nautical miles between 1,046 fixes over 100 days. The satellite tracking buoy had given us a fascinating, but distressing, window into how much a seriously entangled, dying whale could still swim. It was truly heartbreaking to be unable to turn things around for that whale.

Churchill was never seen again. Ever since examining the dead right whale at Newcomb Hollow Beach in 1996, I have thought a lot about the behavior of whale carcasses. Some large whale species, such as right whales, float if in good body condition; others, such as blue or fin whales, sink, as do skinny right whales. Once whale carcasses start to decompose, gas starts to build up, first in the intestines, and then in other tissues and the bloodstream, until they eventually inflate. By this time, all species tend to float. Then, as it decomposes further and tissues start to break down, the carcass often decompresses through the mouth, ejecting liquefied soft tissues and spitting out bones. Sometimes this process leaves a bag of blubber floating on the surface. In the case of Churchill, he was skinny enough that he probably sank once he stopped swimming. Given that he sank in such deep water, the bottom water temperature and the extreme pressure most likely precluded the formation and expansion of enough gas to make his body float again.[13]

In debriefing after Churchill died, we realized that we had started to learn how much sedative a North Atlantic right whale could handle without losing its ability to swim, breathe, and surface. It would be a few years before we would get another chance to develop the system further.

I retreated to my office at WHOI and slowly reintegrated into the life of research, doing science, writing proposals and reports, and publishing papers. During this time, I continued to work with colleagues to examine dead North Atlantic right whales as they were found. My experiences with #2030, Churchill, and a variety of other dead whales had led me to feel increasingly burdened by their suffering and responsible for them as individuals. I led a study that summarized all of the North Atlantic right whale deaths between 1970 and 2002.[14] The species was struggling as a whole, but the more I understood about how individual whales were suffering, the harder it became to document dispassionately what we were finding. My first attempt to go further was at the meeting in San Diego in 2005 that I mentioned at the beginning of chapter 1. I remember standing up at the podium in a large hotel conference room, with my slides, and talking through some of the work we had all been doing. I described how in the previous ten years, 6 of the 64 North Atlantic right whales entangled in fishing gear had died, 35 had been disentangled or shed gear on their own, and 5 more were presumed dead after no sightings for six years. Twelve whales were still carrying potentially life-threatening entanglements. Of entanglements not passively shed between 1999 and 2005, the average duration was at least 10 months.

I explained that chronically entangled whales lose so much weight that they sink when they die, and that therefore gear-induced mortality is underestimated more than ship kills. Often the animal is wrapped up in serial loops of rope (and sometimes net), with gear in the mouth or around the flipper, body, and peduncle. If the gear is anchored to the bottom, has a heavy weight hanging off the animal, or is fixed in more than one place on the body, then as the animal flexes to swim, the wraps of gear cinch increasingly tighter. I described the entanglement of right whale #2030, and how other whales suffered from gear wrapped around their tails for long periods, cutting in to the leading edge of the flukes and peduncle, and eventually into major arteries. There were often substantial amounts of trailing rope causing significant drag as the animal swam. Thus, gear entanglement was a major animal welfare issue as well as being an obvious conservation concern for this critically endangered population.

A gratifying number of people came to hear my talk. The audience asked some good questions and made some nice comments. Then everybody moved on to the next talk, and I was left standing there, alone.

In 2001, we held another workshop at WHOI that was in part a debrief for the attempted Churchill rescue, and in part a discussion about possible technological solutions to the disentanglement problem.[15] We schemed about robotic devices that might attach to and then pull themselves up the trailing rope to cut it from the whale. We discussed devices with suction cups and cameras that might guide a cutting blade. We considered the difficulty of freeing rope from the whale's mouth and armpits, both areas where ropes often constrict and persist.

We then turned back to the issue of sedation. Our experience with Churchill had shown that we were getting close to a workable drug dosage that was beginning to calm the animal and seemingly control pain. We outlined a design that could serve as a drug delivery tool. It had a series of compartments that held drugs to be injected as needed.

We also discussed alternative drug choices, as well as the concern that, once we had reached a working dose that made an animal approachable, the animal might lose equilibrium and balance. This led to a discussion of a variety of physical supports for sedated animals, such as a large inflatable platform that would serve as an at-sea dry dock on which to restrain the animal physically to enable its disentanglement.

There also was debate about whether to pursue disentanglement at all. Some argued that we should focus on preventing entanglements in the first place. Others saw entanglements as unavoidable and disentanglement as a critical stopgap measure. They were both right.

Monitoring approaches we discussed included tracking tag options and use of a behavioral monitoring tag during a sedative-assisted disentanglement approach. This last idea was raised by Teri Rowles. In a few short years, she was to raise it again in a manner that was hard to ignore.

The workshop was followed by a grant from NOAA to the Center for Coastal Studies to work with me at WHOI to develop a ballistic dart system for large whale sedation. My biggest challenge was to find a ballistic dart system that was functional and affordable. I was aware of a company called Paxarms in Timaru, New Zealand, that had made ballistic whale biopsy systems. I had a long phone conversation with Trevor Austin at Paxarms. I summarized the results of the workshop, and he quietly and confidently told me, in his uniquely New Zealander, low-key manner, that he could provide what I was asking for, but that he would need support to design and build the system. I had to decide if I was going to commit to this system, as inevitably our resources were finite. However, Trevor's cost quote, which he delivered in July 2006, was affordable. I consulted with NOAA and CCS, and after meetings and correspondence about the detailed specifications, their advice was to proceed, as there was no other source that I had been able to identify that was competitive with Paxarms. With a certain amount of trepidation, but with an underlying sense of confidence that Trevor would deliver, in September 2006 I signed the purchase order and navigated the institutional red tape.

Questions followed from CCS as to when the system would be delivered, and from Trevor as to how some of the details would be worked out. Were we importing a firearm or not? (Not if it couldn't fire a bullet.) The plan was that Trevor would hand-carry the drug delivery system to us from New Zealand, train us in how to use it, and use the whale blubber I had stockpiled in the freezer to test it. In March 2007, Trevor arrived at Logan Airport in Boston with the system. He and I drove along Route 6 to Provincetown to CCS to start the testing and training process. We worked at a local shooting range in bitter cold. Trevor, despite having left New Zealand at a temperature of 77°F (25°C), survived his plunge into 5°F (−15°C) weather. The system worked well. The only significant problem was that Trevor had been firing darts into hay bales at home. Once he was shooting into whale blubber, the strength of the blubber arrested the needle more immediately on entry than the hay had. The resulting rapid deceleration caused the liquid-filled dart body to fold over the stainless-steel needle tube at the base of the needle. Trevor had been suspicious that this would happen, so he had brought some carbon fiber tubes that could line the needles. Once a tube was glued in place, the characteristics of the needle were far better. The carbon fiber tube allowed the needle to spring back into alignment with no damage, allowing the drug to be injected within 3 seconds of implantation in the blubber and muscle. After we had worked on the system with the CCS folks for a couple of days, Trevor and I returned to my home, where he was able to make some more adjustments to the sights and air pressure control to get the system well set up. Trevor returned home, and then it was a matter of waiting for a suitable case.

On May 16, 2007, a mother and calf humpback whale pair were seen 72 nautical miles inland, in the Sacramento River basin.[16] They both had significant propeller trauma from a large ship. The wound edges were not healthy, and skin color was poor. Attempts to herd the whales down the river were not particularly effective, although over the next twenty days, mostly at night, they moved back down the Sacramento River to the Golden Gate Bridge and the Pacific beyond, never to be seen again. Given their poor condition, I was asked to bring the Paxarms whale dart system to California to help administer antibiotics to both animals. My primary contact at NOAA for the development of the Paxarms system was Jamison Smith. Jamison had grown up hunting, was a good marksman, and understood guns. He and I joined the convoy of boats escorting the animals down the river, and we were able to deliver antibiotics to them using the dart system.

Thus, we were able to show that we had a system that worked on the water and was capable of delivering drugs.

We then waited until January 23, 2009, for an entanglement case in which sedation was appropriate. North Atlantic right whale #3311, born in 2003, was last sighted gear free on April 21, 2008, in Cape Cod Bay. It was first sighted entangled by a crew from Clearwater Marine Aquarium off Brunswick, Georgia, on January 14, 2009, with multiple lengths of line coming out of the left side of its mouth and wrapped around the left flipper. A rope had cut down into the upper aspect of the upper jaw and the lower lip. The crew attached a satellite tracking buoy to the trailing entangling gear that afternoon. The whale was then located off Mayport, Florida, by the next morning, and off Daytona Beach the following day. By January 20, the whale had moved north again, to Cumberland Island, Georgia. On January 22, the crew again attempted to disentangle the whale. Throughout the day, the animal was highly evasive, changing course, delaying surfacing, and diving upon the boat's approach. On January 23, after estimating the weight of the animal, we administered two doses of drugs, but the whale actually swam faster and became more evasive.

The satellite tracking buoy then tracked the animal to just south of Block Island, Rhode Island, on February 16, before it headed south again. We made a further sedation attempt on March 5, off Ponce de Leon Inlet, Florida. The animal's respiration rate seemed to increase, and its breathing seemed shallower. It remained evasive and entangled. We agreed to increase the dose the next day, March 6. As a result of the increased dose, the whale swam faster, but became more approachable. A cut was made in the entanglement, and much of the rope and the satellite tracking buoy were freed from the animal. On previous days, observers in an aerial survey plane had reported that the whale would abort a surfacing if the boat was waiting for it to surface, but this behavior changed on March 6. Thus, it seemed that we had reached a drug mix and dosage that was allowing us to approach and help the animal. It was remarkably humbling to aid in the removal of entangling rope from a whale. Once the animal was freed, its swimming speed increased and its respirations became deeper.

Sadly, from January through March, visual observations of the whale had reported a steady decline in its body condition and health. Its patchy skin discoloration had increased, it looked more emaciated, telltale "rake" marks were appearing around the blowhole, and whale lice were spreading all over the body. All very poor prognostic signs.

As I write this, #3311 has not been seen since March 6, 2009. Eleven years later, we can assume the animal died due to deteriorating health as a result of entanglement, despite our removing much of the rope. Thus, our effort was too little, too late. At the time, of course, we did not know this would be the outcome, but we all had our strong suspicions that the animal was not going to recover, despite our giving it a better chance by removing the rope as we had.

In the fog of war, it is often hard to get a good perspective. We knew that we were only struggling to save individual whales, and probably in vain. Yet we still had no sense of where this was all going to end, and how.

A year later, on December 23, 2010, another candidate for sedation was first sighted entangled off Jacksonville. The case of this female North Atlantic right whale was managed by a combination of teams from the Florida Fish and Wildlife Conservation Commission and the Georgia Department of Natural Resources. They reported that she had rope trailing from her mouth, and that both flippers were involved. Her body condition was poor, and wounds associated with the entanglement appeared to be scarring over, suggesting that her entanglement had been ongoing for some time. The New England Aquarium assigned her a number of #3911. On December 30, the teams cut and removed some of the gear to enhance the chance that the whale could shed the rope, but there was a thick twist of lines exiting the left side of her mouth. They removed a significant amount of trailing gear, including a piece of plastic-covered wire mesh attached to a piece of rope—most likely from a lobster trap, given its appearance.[17]

As the satellite tracking buoy the team had attached reported the changing location of the whale, we watched the weather forecast and the whale for a possible day to attempt both traditional and sedation approaches for freeing the whale. We reviewed the state of the animal, discussed the entanglement, and weighed the risk of sedation. We agreed that attempting to disentangle the animal without sedation was the conservative first step, and only once that approach had failed would we consider use of sedation. Teri Rowles, as always, was asking the hard questions: How can we better document the effect of the drugs on the animal's behavior? How can we track the animal after sedation to monitor its recovery? While neither of these answers was easy, we had access to tags suitable for addressing each concern. Teri gently and persistently led us to a point where we had workable plans.

There were tense discussions, especially in light of a pending snowstorm, as to what date might be workable in terms of travel from New England, the whale's location, and the Florida weather. January 12, 2011, saw Jamison Smith and

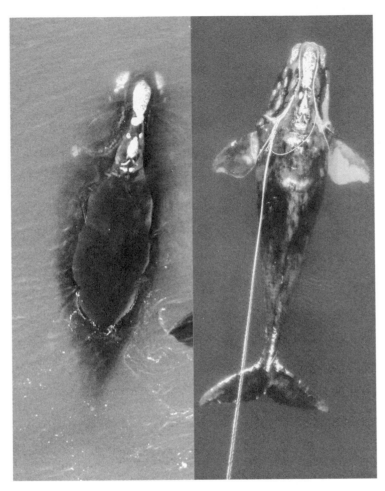

North Atlantic right whale #3911. *Left*: February 20, 2010. *Right*: December 30, 2010. Note how emaciated the whale became with the entanglement. Photos: Florida Fish and Wildlife Conservation Commission. Permit: NMFS #594-1759.

his colleague David Morin digging their cars out of a snow-storm near Gloucester, where they both lived. Airports were closed, so once dug out, they picked me up, and we drove south into a blizzard. January 13 found us in a hotel room in New Smyrna, Florida, right on the beach, with the intent and anticipation of heading out of Ponce Inlet the next morning.

The next day it was windy, and the animal passed us by. We all met and stepped through what we had planned: the details of how the different boats would interact, the steps to take, the priorities, the pitfalls, the unknowns, the hopes and fears. I checked the behavioral monitoring tag system we planned to use, charged its battery, and tried to be sure that all the stuff we needed for the tag and sedation systems was organized and ready to go. One of the biggest challenges in these efforts is the lack of space on a small boat. How much gear should we take, what spares should we leave behind? And do we have the right team? It was hard to hire "at-sea whale sedation experts" who had prior experience. And how exactly will it all pan out? But that is all part of the mystery of this work—the endless mental loops as one tries to think through potential scenarios in the hope of having the right solutions as reality finally unfolds.

At 10:04 a.m. on January 15, Chris Slay, skipper of the disentanglement boat, made a careful approach to the animal, and the monitoring tag was attached in a good position on her body. At 10:15 we began the sedation approaches, and the animal was darted at 10:24. We had all agreed to wait for 30 minutes to allow the drugs to take effect. Our approach to the animal to dart her was the first time I had a chance to examine her, other than in the many photographs that were available from the previous disentanglement attempts. There were chafed areas of abraded skin and blubber around the head, peduncle, and flukes. In a side view, we could see a marked dip in the body outline behind the blowholes, patches of discolored skin, and whale lice. None of this was a surprise, but getting these close, though fleeting, views that showed how sick the animal was made the whole case that much more urgent and distressing. For our boat, it was time to hang back and let the disentanglement boat do its job. It made a number of approaches starting at 10:59, but the whale was still avoiding the boat, as she had on previous disentanglement attempts in January. But by 11:15, the boat was able to get alongside the whale. By 11:33, a cut was made and a trailing rope removed. By 11:53, many lines had been cut, and a minute later, the buoys that the whale had been towing (added during the disentanglement attempt to slow her down) stopped moving, and the whale swam off without them. It was then time to dart the whale with drugs to reverse the sedatives, deliver antibiotics, and attach a new tracking tag. This was all done by 2:00 p.m. We followed the whale, watching her surface and start to show her flukes when she dove. The weather then worsened, and we had to return to port.

It was only once we were back at the dock that I was able to catch up with Chris. As we both sat down on an old piling, I was desperate to know if he thought the drugs had let him approach the whale, or if she had just been more approachable for another reason. While we were on the water, it had been hard for me to tell what was going on once we had injected the sedative and dropped back to let Chris and his team do what they did. He looked at me, smiled, and confirmed that indeed, it was very clear that as the drugs had taken effect, the animal was far less disturbed by the boat's approach. When discussing the case more recently, in an email of July 30, 2020, he said, "I like your characterization, except for perhaps the world 'let.' Humpbacks 'let' you disentangle them. This little right whale didn't let us, so much as it seemed unable to react in the typical side-winding right whale fashion, clearly being under the effects of the dose administered. It simply swam along, seemingly unable to muster the typical erratic reaction, not unlike a counter-espionage agent, drugged and being walked by the elbow into a waiting van."

We made it to Jacksonville that night before we fell into our respective deep sleeps in a motel, then headed home the next day. I spent the following day in the lab, examining the used darts that we had retrieved from the whale, trying to learn as much as I could about how they had operated and what we might do to make the process as safe and efficient as possible.

7

Our Skinny Friend

Right whale #2030 had driven me to pursue the idea of inter- vening with entangled large whales at sea with the hope of alleviating their suffering and enabling them to become productive whales again. Ten years later, #3911's case was another turning point, when I came to realize that such an approach, while a good thing to do, was never going to be enough. The problem was rope, and we could, and did, modify how rope was used. But as long as rope remained in the water column, the problem of entanglement would remain. The US government's management approach was to reengineer the way rope was deployed to stop it from being lethal. It was a laudable goal, but never achieved. Furthermore, with the majority of North Atlantic right whales facing sublethal entanglement trauma, the species wasn't going to grow. The sublethal trauma was degrading their health, making it harder for them to reproduce. Thus, we needed to examine the sublethal impacts of entanglement to better understand the extent to which they were a problem, and to evaluate whether actually getting rid of rope, as opposed to modifying how it was still used, was essential.

. . .

On my daily commute to work, I continued to ruminate on the partial disentanglement of right whale #3911. Whales were still getting entangled in fishing gear, but with drug delivery, we had calmed a large whale and seemed to help it. The obvious question was whether we were too late: had #3911's health deteriorated too far for her to recover? We also did not know how much rope remained in her mouth, and perhaps wrapped around the base of her flippers, that we could not see. On January 17, I wrote a debrief email, signing off with, "Here's to our skinny friend making it back north to the biggest swarm of *Calanus finmarchicus* any North Atlantic right whale has ever had the chance to consume [*Calanus* being the species of copepod that North Atlantic right whales prefer to eat]. I think it is still touch and go for this case."

On February 1, just two weeks after we had sedated #3911, I received a call and learned that she was dead. It was with a heavy heart that I shared the news with Julie van der Hoop, who had been studying whale entanglement in my lab that fall and into 2011. We had talked about the whale before I had gone south the previous week. I told her we would be leaving for Florida as soon as we had grabbed the necropsy gear we would need. She looked at me with deep concern, and we were off to the airport.

An aerial survey plane had sighted the carcass of right whale #3911. She was towed to the beach the following day. Julie and I arrived as the beach team was starting to haul the whale up the beach. I had undertaken the same task in the past, but the folks in charge were making good progress. It was good to wait and watch. A large cable broke. Tom Pitchford, from the Florida Fish and Wildlife Conservation Commission, was in charge. It was Tom who had been driving

Peduncle and flukes of North Atlantic right whale #3911 on December 25, 2010, showing massive trauma and scarring from abrasion by rope trailing from the mouth and flipper. Photo: Florida Fish and Wildlife Conservation Commission. Permit: NMFS #594-1759.

the sedation boat two weeks previously. I was waiting to see what would happen next. Sure enough, he quietly walked back up the beach to his gear trailer, opened the door, and emerged with a long heavy-duty tow rope to finish the job. I had bought a set of those ropes a few years previously with the government support I was working under, and had left one with Tom. He had taken care of it, and it was ready and waiting. It was good to see.

By this point, we had all done far too many North Atlantic right whale necropsies, and we had established a comprehensive examination and sampling protocol.[1] For entanglement cases, our analysis synthesized what we knew of the case before the animal died, in terms of the duration and nature of the trauma as observed at sea; postmortem trauma; measurements; and documentation of any fishing gear removed from the whale, which was then archived. These data, along with the previous sighting record for the whale when alive, and any other relevant analyses, were then summarized in a case report. These reports have become important documents in the heated legal, political, industrial, and conservation stakeholder interactions that have continued as the right whale species has become increasingly imperiled in the last decade. However, an understanding of how the rope and whale behave at the start of an entanglement is still elusive.

I was reluctant to lead this necropsy. As part of the sedation team, I was no longer an impartial observer. But I really wanted to examine the locations where we had injected the animal, and to learn how much rope remained in the mouth or elsewhere. The day-long necropsy was tough. The team was led by Bill McLellan, and Alex Costidis, from the Virginia Aquarium, both skilled in large whale anatomy and forensics.

Right whale #3911 was 33 feet (10 m) long. The ends of multiple ropes that had been cut during disentanglement attempts over the previous two months protruded from the baleen. Each rope was tagged and numbered before it was removed. Rope overlying a scarred furrow ran from the base of the right flipper to the corner of the mouth. Deep shark bites were found on the underside of the tail and the edges of the flukes. Once we got to see inside the mouth, we found loops of rope wedged between baleen plates and hanging in the mouth cavity. Furrows with the impression of stranded rope had been pressed into the tongue. Perhaps the most upsetting for me was the rope scar we found running along the upper edge of the right lip. The rope was still there, buried in the flesh. Cutting through that area, we could trace where the rope had sawed down from the skin surface, leaving an entry point now marked by thickened, scarred skin—repair tissue that had subsequently grown over the rope. We would never have been able to fully disentangle that whale without the capacity to immobilize her with anesthesia and undertake the necessary surgery—a hopeless proposition.

Examination of the severed blood vessels associated with the shark bites around the tail showed tissue fluid accumulation around the vessels, evidence that the animal had been alive at the time of the bites. The conclusion of the necropsy was that the proximate cause of death—that which started the demise of the animal—was entanglement in rope, and that she ultimately died of shark bites.[2] Weakened by the chronic entanglement, she was easy prey.

We then rolled the carcass onto the belly, using a large excavator, and my heart stopped beating for a second or two. We had been able to retrieve two of the four needles we had injected into the animal, but two were missing. One was now staring me in the face, protruding out of the left flank of the animal where I had shot her. I took myself away from the carcass to gain a semblance of composure. In my attempt to help the animal, I had caused her more pain. The dart needle had not changed the whale's prognosis, but I was saddened, embarrassed, and chagrined.

At that moment, I promised the spirit of this poor animal that had suffered at the hands of humans, including my own, that I would not rest until whales were no longer subjected to such horrendous traumas, whether inflicted by rope or by needle.

Alex dissected the area around the needle. It had been designed to pass through the blubber into the muscle, and so that the drag from the animal swimming through the water would then pull on the syringe barrel and extract the needle. This had happened every time we had used the system up to that point. But in this case, the needle had an 80-degree bend at the junction of the muscle and blubber. Effectively, it was a large barb. There was no way the water drag on the syringe barrel would have pulled it out as designed. Further-more, the needle tip had carved a cavity in the muscle.[3] We subsequently showed this to have resulted from the rhythmic shearing of the blubber relative to the muscle every time the animal raised and lowered her flukes to swim.[4] This shearing is most pronounced farther back in the animal. The needle should have been placed farther forward.

The behavioral monitoring tag we had attached prior to sedating #3911 was a digital acoustic recording tag (DTAG), a pressure-proof package the size of a smartphone that syn-chronously records the animal's orientation, movement, and emitted and received sounds.[5] We wanted to document the impact of the entanglement on the ability of the animal to swim and dive, the effects of sedation on its movement, and hopefully, the change in its behavior after it was disen-tangled. The tag is attached using a long pole to place it on the back of the animal, to which it adheres by means of suc-tion cups, and programmed to release itself from the animal at a given time and float to the surface, where it is relocated by radio tracking. Once the tag is recovered, the data are downloaded and analyzed.

Over the following months, I was keen to get the DTAG data analyzed to start testing the hypothesis that entangle-ment drag has a significant energetic cost to North Atlantic right whales. Julie van der Hoop expressed strong interest in those data, as no one had ever had a look at the fine-scale movements of an entangled whale—nor, indeed, of a sedated whale—at that time. Julie had a growing interest in the physiological effects of entanglement drag on large whales. She entered the MIT-WHOI Joint Program in Bio-logical Oceanography that June, and one of her first projects was to tease apart the complex DTAG data set. Despite never having done so before, she made short work of getting a useful analysis from the #3911 data. The study she was able to do was based on the DTAG data covering six hours of swimming in four phases: entangled, entangled with the onset of sedation, sedated but disentangled, and disentangled as the sedation wore off.[6] Following the removal of the entangling rope, she found significant increases in the whale's dive depth and duration; increases in ascent, descent, and fluke stroke rates; and decreases in fluke thrust. The data showed that entangled whales can have significantly increased energetic demands, whereas sedatives have more subtle effects on their swimming behavior.

To put these findings into a better context, we needed to calibrate the drag that came from towing a defined amount of rope through the water at different speeds relevant to a swimming right whale. For that, we needed a suitable boat. In 1987, three years after Hannah and I got married, we had moved back into the home she was raised in, as her father had died a few years earlier and her mother had moved to Florida. The house is on a small island in Buzzards Bay. Islands beget boats, so there have been a number of instances when I have turned a small boat into a research platform. In 2003, for example, we had funds to develop a system to lasso a whale by the tail to assist with disentanglement. We had a hollow fiberglass, full-scale, lifelike replica of a North Atlantic right whale's flukes that we could fill with enough water to make it barely float and tow behind a small boat. Increasing the speed of the boat made the tail tilt and submerge, while slowing down made it surface again, in such a way that it would look like it was swimming. The idea was that the tail would help disentanglement teams develop and practice their techniques. Two colleagues would follow behind in another boat to test the converted net gun that was to throw the lasso over the flukes. After a few trials, in which we successfully caught the "whale," the then-harbor master came out in his boat to investigate a report of a small boat shooting at a live whale that was swimming in the inner harbor—apparently the model was quite realistic.

May 2011 found Julie and me using the same boat to tow the rope that had been removed from #3911 through the water. A tension meter placed between the boat and the rope gave us a continual record of how the drag force changed with the speed of the boat. Julie then combined those data with the DTAG data. Her paper showed that entanglement could increase the power required to achieve a given speed through the water by 60–164 percent, depending on the amount and configuration of the gear involved.[7] She had shown us that the energetic cost of entanglement could be substantial. This finding would not surprise anyone who has tried to hold onto a significant amount of rope being towed behind their boat.

What was that energy drain doing to a North Atlantic right whale's ability to get pregnant, suckle, and wean a calf? This basic question was the driver for a major portion of Julie's subsequent doctoral dissertation. Just as dairy farmers run a business that depends on feeding their cows to derive a profitable milk yield, North Atlantic right whales need a net gain in energy sufficient to make new baby whales, and to feed them in utero and while suckling, so they can recruit those babies into the family trade, thus growing the North Atlantic right whale species business. Energy drain from entanglement drag can be compared to intestinal parasites sucking the energy out of their host. Any dairy farmer knows that cows with high parasite loads do not do well.

In September 2012, Julie drove down to the Narraganset, Rhode Island, warehouse where the government held an archive of the fishing gear removed from whales during

US disentanglement operations. This archive was a unique resource, in that a timeline of known events related to each entanglement case, the visible impacts of the fishing gear, and the outcome were recorded for each set of gear. Julie carefully documented the nature and dimensions of the gear and of any traps or floats associated with it. She then designed a system to tow it behind a larger vessel in deeper water than we had previously worked in. Over a long day, we were able to tow fifteen different sets of gear, at a range of speeds and depths, again using a tension meter to record the resultant drag forces. She concluded that combining gear drag measurements with theoretical estimates of drag on whales' bodies suggested that, on average, entanglement increases drag, and thus required propulsive power, by a factor of 1.5.[8] For a human, that factor would be the difference between light and heavy physical labor. That study was the first of a series that she subsequently published, explaining what gear drag was doing to these animals' ability to raise calves successfully. I remember discussing with Julie the evening after our dragging trip how the data we collected that day could become the foundation of her dissertation.

One of the cases that Julie tested was that of North Atlantic right whale #4057, a 3-year-old male that was found entangled on February 16, 2014, off Jacksonville, Florida. His most recent pre-entanglement sighting was eleven months earlier, on March 18, 2013, in Cape Cod Bay. His body condition was fair. He was entangled in 500 feet of three-strand synthetic rope entering and exiting the left side of the mouth, with one end just above the eye and the other end trailing more than 100 feet behind the flukes before sinking out of view. Prior to the disentanglement effort, researchers had attached a DTAG to his left flank. Thus, we had two entangled North Atlantic right whales that had been tagged during disentanglement. Julie was able to take a close look at how entanglement drag affected the whales' gait, thrust, and fluke movements, getting a better understanding of the problems these animals faced when entangled and how they compensated for the added drag. Removal of the entangling rope reduced the amount of thrust that the whales had to generate to move at a given speed. So, the whales were adjusting to their increased load, just as humans do when cycling in a head wind[9]—the major difference being that when humans tire of their load, they can shed it at will. The whales are stuck with it until they can eventually shed it, a human comes along and removes it, or they die months later.

Julie and I spent many, many hours discussing her data, what they all meant, and how to best shape her findings into an objective assessment of the broader implications for North Atlantic right whale health, fecundity, and survival. Ultimately, she fleshed out the concept of entanglement in the context of the whale's ability to live and make more North Atlantic right whales.[10] To get at this issue, she took two approaches: first, she compared the blubber thickness of lethally entangled whales with that of healthy whales to estimate how much blubber the dead whales had lost. Second, she calculated the energy requirements of swimming with a given amount of gear drag, concluding that an average entanglement lengthened the reproductive cycle of a female North Atlantic right whale by up to 8 percent, delaying calving by months to years. Thus, entanglement can be a costly life history stage that impedes a whale's ability to get pregnant and wean calves. By the numbers, she showed that an average entanglement could drain the energy equivalent of that required for suckling a calf, and that even relatively minor entanglements could delay an animal's ability to get pregnant. Thus, entanglements not only caused deaths, but also made it harder for the species to grow.

At the same time, Amy Knowlton at the New England Aquarium undertook a massive review of all the photographs in the North Atlantic right whale catalog.[11] She found that 83 percent of the catalogued animals carried entanglement scars, 59 percent had been entangled more than once, and 26 percent acquired new scars every year. Combining her data with ours showed that entanglement is not only a cause of death, but also has a major impact on the health of living whales.

How many more female North Atlantic right whales would have calved if they weren't entangled? This question is a complex one that needs to be considered in the context of all the relevant parameters, which include seasonal and spatial variation in feeding success; the impacts of sublethal entanglement stress, and the effects of other stressors, such as episodic and persistent anthropogenic noise. They may also include other poorly understood factors such as the effects of ingested plastics and toxicant impacts, due to both exposure and release from body stores in times of negative energy balance. For instance, lactating mothers would pass released toxicants on to their young through their milk, and would also be exposed all over again as the chemicals circulated in their bloodstream.

Since Julie left my lab, I have been fussing over the question of how entanglement stress interacts with foraging success and with various other stressors. Whenever I saw Julie, we would talk about it. The whiteboard in my office sported an evolving right whale energy budget, with income, expenses, and uncertainties. Then, in early 2019, I received an email from a student at the University of Kiel in Germany, looking to study in my lab for her undergraduate thesis. It took a while for things to line up, but on March 1, 2020, Jasmin Hütt duly arrived. We stared at the whiteboard for a while, talked about process, and consulted with Julie, as well as with Peter Corkeron at the New England Aquarium, a marine mammal modeler that I respect hugely. Jasmin started to read the mountain of publications I steered her toward, and many more besides. Ten days later, WHOI closed due to the COVID-19 virus pandemic. A few days after that, we realized she would be best off at home. The rest of her thesis work happened remotely. After a few calls with Julie and Peter and some elegant thesis drafts, Jasmin was able to build a comprehensive model of the food energy income and the migratory, metabolic, and other expenses normal to a right whale, and overlay them with varying levels of food availability and entanglement stress. In so doing, she was able to model the growth of the inter-calving interval in recent years, from the normal 3-year cycle, to three times that, primarily due to reduced food availability as a result of climate-driven water temperature shifts, but also due to entanglement stress.

Over the preceding four years, I had been undergoing a slow, complex metamorphosis. In April 2016, I had the good fortune to be given a new kidney by a close friend and colleague, Julika Wocial. She had worked in my lab in 1999, helping with Carolyn Miller's studies in the Bay of Fundy, and we had stayed in touch. I have a genetic kidney disease that finally required a transplant to avoid the imminent need for dialysis. A number of colleagues, family, and friends were willing to be tested, but Julika drew the short straw. The transplant changed my life in fundamental ways. I had six months of post-transplant recuperation away from work, during which I contemplated my past work, my incredible fortune in the gift from Julika, and what I was fit for now that I was immuno-suppressed due to the anti-rejection drugs that I would need for the rest of my life. Obviously, my habit of full-body immersion in rotting whale carcasses, and their associated bacterial soup, might be unwise to continue. That wasn't easy to accept, but I did so because of the number of people I had worked with over the years who had become better at the job than I was, and because I realized that stepping back was a good thing. It was also obvious that we knew more than enough about the major stressors that were an existential threat to the North Atlantic right whale species, so that I could usefully transition away from beach necropsies toward a more active role in reducing entanglements and vessel strikes. This possibility had been a cloud in my brain for a long time. I really enjoyed the diagnostic process, but I also knew that prophylaxis had to be the ultimate goal of any veterinarian. Yet I had resisted becoming part of that endgame.

But the time had come for me to get more involved in the politics, economics, and realities necessary to change whales' odds of lethal and sublethal vessel strike and entanglement trauma. Reducing trauma from both sources is critical. Vessels have indeed been subjected to speed restrictions and excluded from certain areas. But right whales and other large whale species continue to be affected by vessel trauma. The areas that are managed are far too small, as the collision risk is widespread. The entanglement problem for right whales is worse than ever. The final chapter of this book, chapter 8, will take a nuanced view of how we can resolve both issues to secure coastal and offshore waters for large whales and human commerce in the centuries and millennia to come.

8

Taking the Long View

WHY CAN'T WE LET RIGHT WHALES
DIE OF OLD AGE?

There were once tens of thousands of North Atlantic right whales.[1] The vast majority were killed by harpoon whaling over more than a thousand years. By 1990, approximately 270 remained. By 2010, the population had slowly grown to about 482. By 2020, it fell again to an estimated 355.[2]

Today there are many more fishermen, managers, right whale conservation biologists, and lawyers concerned about, and committed to changing the critically endangered status of, the North Atlantic right whale, than there are such whales. These people constitute a strong, collaborative community. The story I tell in this book is just one story within the history of that larger group. There are many others who have dedicated far more of their lives than I have to understanding, and hence conserving, the species. Yet it is not enough: if it were easy to fix this crisis, we all would have made it happen by now. Just as it took more than entrenched scientists to enact a commercial whaling moratorium, the right whale needs an awakening of many more people willing to understand and to act.[3]

The intent of this chapter is to show that climate change–driven ocean instability, with consequent right whale movement unpredictability, has made entanglement and vessel trauma even harder to manage. This change has invalidated the current paradigm of North Atlantic right whale conservation by management of vessel strike and entanglement hotspots. We need a broader, more generic approach to minimize the risk of trauma to whales. Such an approach will succeed only if there is a broad-based consumer demand for commodities and seafood that are transported and procured in a manner that allows whales such as the North Atlantic right whale to swim free and prosper. Furthermore, an attitude of ethical consumerism would have benefits to the conservation of global biodiversity far beyond the North Atlantic right whale.

There is an instructive success story from the turn of the twenty-first century in eastern Canada, but the success lasted only while the whales continued to behave as they had in the past. At that time, the whale conservation effort focused on reducing the risk of vessel collisions, the primary source of mortality then. It became clear that we needed to know where the major collisions were occurring and to reroute ships away from those areas. Where rerouting was not possible, we needed to slow ships down, because slower ships are less likely to cause lethal injuries (although even a ship traveling at 10 knots has roughly a 30 percent chance of inflicting a lethal injury if it strikes a whale). Slower ship speeds might give whales a chance to get out of the way if they detect the vessel and make an evasive response. This solution might seem simple enough, but to make it happen, we needed collaboration—between the shipping industry, nongovernmental organizations, relevant governments, and in some areas, the International Maritime Organization, the "global standard-setting authority for the safety, security and environmental performance of international shipping."[4]

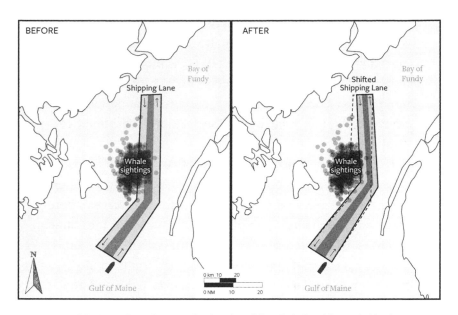

Map of the Bay of Fundy, Canada, showing right whale densities and shipping lanes before (*left*) and after (*right*) the lanes were moved to the east to reduce vessel/whale collision risk. Credit: Modified from plot by Kerry Lagueux, New England Aquarium; © Woods Hole Oceanographic Institution, Natalie Renier, WHOI Creative.

This collaboration first focused on the Bay of Fundy, where the Canadian government had already delineated a North Atlantic right whale conservation area. A group of mostly Canadian scientists, spearheaded by Moira Brown of the Canadian Whale Institute and Chris Taggart at Dalhousie University, took a long look at North Atlantic right whale habitat and the paths of vessels transiting the bay to deliver oil and other goods at the ports of Saint John and Bayside, New Brunswick, and Eastport, Maine. Using data on North Atlantic right whale sightings and ship traffic in the area, they estimated the risk of a lethal vessel strike in the Bay of Fundy shipping lanes and the surrounding vicinity. They found a risk hotspot where the conservation area intersected with the outbound shipping lane to the east of Grand Manan Island. The scientists then repeated their analysis with the shipping lanes moved just a few miles to the east. Their models predicted a 62 percent reduction in lethal collision risk if the lanes were moved, which occurred on July 1, 2003, after adoption of the amended shipping route by the International Maritime Organization.[5]

Furthermore, to mitigate the risk of vessels colliding with whales southeast of Cape Sable, Nova Scotia, researchers undertook a study in that area to determine the places where there was the greatest risk of vessel trauma, and identified Roseway Basin, about 40 miles southeast of Cape Sable. Roseway Basin had long been known as a right whale area by Canadian whalers hunting other species in the mid-1960s and by Fisheries and Oceans Canada, which made it a conservation area in 1993. The Canadian government proposed to the International Maritime Organization that Roseway Basin be made a seasonal voluntary "Area to Be Avoided," which was done on May 1, 2008. A later analysis showed that this policy reduced the probability of a vessel hitting a whale from once every 1–2 years to once every 40 years.[6] University scientists monitored ship tracks and worked with the Canadian Whale Institute to send letters to each vessel operator, thanking those who abided by the advisory and sending educational updates to those who did not. Compliance resulted in an 80 percent reduction in collision risk to whales.[7]

The paradigm of reducing vessel-whale collisions by moving shipping lanes and slowing ships has been subsequently used around the world: in waters off the US east and west coasts, in the Mediterranean Sea, and in the Hauraki Gulf, New Zealand. The two biggest challenges have been vessel compliance with recommendations and regulations, and the need to revisit these conservation measures as the animals react to dynamic environmental conditions by changing their behavior and movements. Such changes, in turn, lead to new hotspots of risk, and new challenges for management action, compliance, and monitoring.

In this context, the Bay of Fundy shipping lane adjustment became substantially less effective once many right whales moved on from that area as a result of climate change-induced movement of their prey. Starting in 2012, the number of North Atlantic right whales sighted, and their residency time, in the Bay of Fundy began to drop. As researchers began additional surveys, it became apparent that North Atlantic right whales had started to spend significantly more time in the Gulf of St. Lawrence, probably because warmer waters in the Gulf of Maine and the Bay of Fundy led to a reduction in zooplankton food resources for the whales in those areas.[8] Before that time, right whale sightings in the Gulf of St. Lawrence had not been nearly as common as in those waters to the south. Sighting effort had also been lower, but the finding was nonetheless surprising, as there had also been only two recorded whale deaths in the Gulf; one whale had washed ashore on the Port au Port peninsula on the west coast of Newfoundland in 1988, and another, entangled in heavy rope, stranded on the Magdalene Islands in 2001.[9] Two deaths did not make the Gulf a risk hotspot for North Atlantic right whales. Then, in 2015, three dead North Atlantic right whales appeared in the Gulf. In 2017, there were twelve, and in 2019, another nine. The Canadian government had to scramble to put mitigation measures in place, as both vessel collisions and entanglement, primarily in snow crab trap end-lines, have become major problems in that area.

As environmental conditions change, with consequent shifts in the timing and location of optimal food availability, right whales change where and when they are to be found.

Sightings of North Atlantic right whales in the Gulfs of Maine and St. Lawrence. *Left*: November 16, 2005–November 15, 2009. *Right*: November 16, 2015–November 15, 2019. These sightings have not been corrected for sighting effort. Nonetheless, the shift to the Gulf of St. Lawrence is obvious, and the reduction in sightings in the Bay of Fundy since 2015 is well documented. Source: NMFS NEFSC, Woods Hole, MA.

Thus, conservation measures that have been implemented for an earlier scenario may no longer be relevant. The shift of shipping lanes to the east in the Bay of Fundy, for example, is no longer relevant to animals that are now going elsewhere— only to the very few that are still using that area. Conservation of endangered large whales is an endless cycle of recognizing a problem, documenting it, designing and implementing solutions, and monitoring their effectiveness. Then, as the parameters change in terms of where and how the stressors and the affected animals interact, another turn of the cycle must begin.

The North Atlantic right whale is currently facing major challenges in Canadian waters, and it is unclear if the species will survive those new threats along with the ongoing sublethal and lethal trauma in US waters. However, the Canadian government has been using an adaptive management approach, revisiting and adjusting vessel strike and entanglement mitigation measures each year since they were implemented in 2017, and making those modifications within the year in which mortality is observed. This adaptive mitigation approach is not just the "implement and maybe monitor to see how things are going" strategy that has been used in the past, but actually aims to improve mitigation on the fly as information and carcasses pile up. Vessel compliance with mandatory mitigation measures in the Gulf of St. Lawrence is monitored 24/7 by the Canadian government, and noncompliance cases are investigated and fined. Whether this approach is adequate to avoid repeating the catastrophic years of 2017 and 2019 is unclear at this point, but the regulatory framework in Canada does have the capacity to make change happen quickly.

In the United States, such flexibility is largely absent due to poorly enforced laws and legally mandated provisions for multi-step public review of any possible actions, which can result in years of deliberation, frequent challenges in court, and dependence on judicial precedent. There has been little adaptation of US vessel strike mitigation strategy in at least ten years, since speed restrictions were introduced in selected areas during certain seasons (Seasonal Management Areas), despite several vessel strike deaths outside of those areas in time or space. Simple modifications could have been made as more has been learned, but these new data have not resulted in adaptation of existing restrictions to make them more effective, although there is a system by which voluntary speed restrictions can be requested when aggregations of right whales are seen in particular areas. Any sense that the

US right whale–vessel collision problem had been solved was shattered in 2020, when two of the ten calves born that year were killed by vessel strikes. This led me to tweet:

> 20% (2/10) of 2020's N Atlantic right whale calves have died. Since 2017, 41 of the ~400 animals left have died: 10% of the ENTIRE species.[10]

How do we manage the generic, persistent problems of vessel collision and entanglement when the time and place of the most serious risks can change in patterns that we increasingly cannot predict as climate-driven environmental change accelerates? We have to understand the generic solutions and apply them widely enough to encompass the uncertainty of where and when the risk is greatest at any one time.

In addition to the failure of the United States to manage vessel collision risk in time and space, its efforts overall to modify fishing practices to reduce whale entanglement trauma have remained mired in conflict. The need to conserve commercially important harvests of fish and shellfish stocks, such as cod, haddock, and lobster, has outweighed whale safety. Compromise—in response to political, legal, and economic pressures—has left the North Atlantic right whale species highly vulnerable. The idea (popular in some circles) that right whale entanglement is now solely a Canadian problem is not supported by the data, wherein 11 entanglement deaths or serious injuries were reported (with varying levels of confidence) in US waters between 2017 and 2020.[11] It is highly unlikely that all those entanglements originated in Canada.

In 1994, the US government introduced "Take Reduction Teams," established under the Marine Mammal Protection Act, to protect populations of marine mammals listed as threatened and endangered under the Endangered Species Act. The enabling legislation included a formula for calculating a mortality limit for species survival. For North Atlantic right whales, this limit has remained close to zero. In other words, if the species is to persist, no more whales can die. Congress required compliance—that animal deaths not exceed this limit—within six months of the legislation's passage. Twenty-six years later, in 2020, we have still not eliminated human-caused right whale mortality. North Atlantic right whales tend not to die of old age.

Furthermore, US government whale population assessment has until very recently been shortsighted in its sole focus on known mortality. A more holistic, and beneficial, assessment would also include modeling of deaths not recorded. Variations in currents, carcass buoyancy, water depth, and water temperature can lead to failure to discover whale deaths at sea.[12] This is especially true for chronically entangled right whales, which are likely to sink, having burned up their buoyant oil reserves. It was recently estimated that the current assessment methods undercount entanglement mortality by almost twofold.[13] Models to better establish the extent of this underestimate are being developed.

The US assessment process also fails to adequately consider entanglement and vessel trauma that does not kill animals, but still affects breeding success.[14] It simply documents a dismal calving rate without addressing the causes. As long as we are simply trying to reduce known mortality, while ignoring cryptic deaths and the sublethal trauma that constrains calf production, our efforts are doomed. This is bad news not only for the whales, but also for the industries paying dearly to follow regulations that they are assured will solve the problem. For unless calf productivity is also increased, there will be no recovery, and all of our mortality reduction efforts, while necessary, will be insufficient.

The struggle to mitigate entanglement of North Atlantic right whales in rope through modifications of fishing gear has been ongoing for more than twenty-five years, but increasingly, these efforts appear to be hopeless. Whatever the gear modification solution of the moment, it seemed to be based on a false premise and did not deal with the fundamental problem: *wherever rope and whales coexist, there is entanglement risk.* The vagaries of food distribution, climate change, and resultant adaptations by humans and whales mean that the only fundamentally viable strategy is to remove rope from the water column over broad swaths of whale habitat. We have to do this in cooperation with the relevant industries to ensure that the necessary regulations and new solutions are safe, viable, and affordable, involving the managers and regulators of the fisheries and the whale biologists who understand the risk of inadequate trauma reduction. It is insufficient to say that "our rope isn't the problem," in that any rope in water where whales go is a risk to them. We have failed miserably to predict, or even document, where entanglements actually occur. Therefore, we have to reduce all rope in whale habitats. But this doesn't mean that we need to stop fishing.

Typically, lobsters are caught in a baited trap that sits on the ocean bottom. Close inshore, these traps are often set as "singles," with a retrieval rope, called an endline or buoyline, attached to each trap and to a surface marker buoy. Farther offshore, five to thirty traps, collectively called a "trawl," are often linked together by a groundline, with endlines and surface marker buoys attached to the first and last traps of the trawl. In both cases, the traps are hauled to the surface for retrieving the catch and renewing the bait. The marker buoy has two purposes: to suspend the endline, and to mark the location of the traps. The buoys have marks and colors that identify the permitted owner of the gear, and they communicate to others where there is rope in the water and gear on the bottom.

A lawsuit filed in 1996 by Max Strahan forced the Commonwealth of Massachusetts to reduce entanglement risk in state waters. To do so, it reduced the number of endlines in Cape Cod Bay, a significant winter and spring feeding habitat for North Atlantic right whales, by requiring lobster traps to be in trawls of four traps each. Two other gear modification measures, sinking groundlines and weak links, were also proposed. Traditionally, the groundlines connecting traps were buoyant, so that they would not get tangled up in and chafed on rocks on the seafloor. There was consideration of making groundlines lie on the seafloor to reduce entanglement risk, although the industry was concerned about excessive chafe of groundlines. Another proposal was that so-called weak links, designed to break at a given tension, be placed between the endline and the marker buoy, in hope that if a whale were to be entangled, it could break out at that point. Longer trawls and sinking groundlines were introduced in January 1997. This focus on gear modification, with no outright closures of areas to fishing, reflected the obvious desire of the industry to maintain access to its fishing grounds. Additionally, the regulators hoped that whale disentanglement and removal of derelict gear from the ocean would help reduce North Atlantic right whale mortality to below the legislative limit. None of these measures was sufficient.[15] Twenty-four years later, Strahan had yet to be satisfied, and rightly so. In 2020, a federal judge agreed with another suit he filed, arguing that the Commonwealth of Massachusetts was violating the Endangered Species Act and

Frames from a computer model simulating how right whales get entangled in endlines attached to lobster and crab traps. Credit: Lawrence Howle, Duke University.

had three months to remediate the situation. Sadly, this decision is probably yet another step down a sorry path of endless procrastination.

If gear modification is ever to reduce entanglement, we need a better understanding of how whales get entangled, and a means to test the efficacy of proposed changes. Colleagues have recently built a virtual right whale entanglement simulator that has been highly informative, but actual evidence as to how entanglements happen remains elusive.[16] We still lack any method to test proposed gear modifications on live whales, other than trial and error, which is a hopeless proposition for the whales and the fishermen.

An example of these endless frustrations is the weak link.

For fishermen, the only acceptable location for a weak link was where the marker buoy attached to the endline. The rationale for this solution was the highly dubious assumption that rope entangled in a whale's mouth, or around a flipper or tail, would slide along to the weak link, which would duly break, and the whale would swim free. There is little evidence to suggest that this assumption was valid.[17] When an entangled humpback whale was recently found east of Chatham, Massachusetts, there was rope in the whale's mouth, attached via a weak link to the surface marker buoy. The link and the buoy rested on the whale's head. The weak link had failed to make a difference, once again.

Today there is also a major push to introduce weak rope—rope that is strong enough to haul traps, but weak enough to avoid killing an entangled adult right whale. This solution might help reduce mortality and sublethal impacts, but again could be tested only by trial and error. Whether multiple wraps of weak rope around one or more body parts would enhance the overall strength of the entanglement configuration is unknown. And this solution would be of little help to calves and juveniles lacking the strength to break even the weak rope.

There was also a major effort to reduce rope in the water column in 2009 by requiring that sinking groundlines be widely used with lobster traps in federal and most state waters. This change does appear to have reduced entanglement in groundlines, but a huge amount of rope remains in the form of endlines running from traps to the surface. It has been estimated that there can be up to 920,500 endlines in the waters of the US Northeast region, of which 912,300 are used to catch lobsters.[18]

The residual risk of sinking groundlines to whales is important to know and hard to quantify. In theory, comparing the amount of groundline found on entangled whales before and after 2009, when it was required to be sinking line, should demonstrate the risk reduction. However, there are too many unknowns and biases to accurately quantify the risk in terms of gear in hand. Such gear becomes available for analysis after it has been removed from a live whale during a disentanglement effort, or a dead one. Groundline is harder to identify than endline, in that the latter is likely to have a buoy attached. Furthermore, groundlines weighed down by a trawl of traps are far heavier than endlines, and so are less likely to be recovered during a disentanglement effort. Additionally, if both endline and groundline are recovered, it may be unclear whether the entanglement started in the endline or in the groundline. Thus, given these potential biases and the sparse available data, it is hard to be clear about the residual risk of whale entanglement in sinking groundline, although the available data suggest that perhaps 10 percent of post-2009 entanglements were in groundline and 90 percent in endline.[19] This issue is a fundamental one, as technological developments to avoid use of endlines, which I discuss below, remain dependent on the ongoing use of sinking groundline where multiple traps are fished in a trawl. An alternative perspective on sinking groundline risk could be gained from analysis of where right whale prey is found in the water column in space and time, and the extent to which right whales are close to the bottom when they are not feeding. Entanglement risk in sinking groundline will be greatest when optimal feeding is near the bottom. The best way to avoid any entanglement in trap gear would be to use neither groundlines nor endlines—a laughable proposition from a lobsterman's perspective, but just fine from the whale's viewpoint.

Another whale entanglement risk is gill nets, as I described in chapter 3. They catch not only fish, but also whales, which may run into the net and, after wrapping the net around their bodies, be left with a persistent entanglement, as in the case of #2030.

By 1997, given the continuing lack of effective management changes, NMFS introduced a series of closures to gill net fishing in Cape Cod Bay and the Great South Channel, southeast of Nantucket, and further restrictions on lobster fishing in the Great South Channel. Both areas had been designated as critical habitats for North Atlantic right whales. NMFS abandoned further gear modification proposals when the industry complained. Nonetheless, the government concluded that what it had done would be sufficient to reach the legally mandated mortality limit. It was wrong. Twenty-two years later, in 2019, the Conservation Law Foundation won a lawsuit reversing a recent decision by NMFS to open an area south of Nantucket to gill nets.

Despite an increase in funding for North Atlantic right whale recovery early in the first decade of the twentieth century, now-retired US Marine Mammal Commission policy analyst David Laist, in his 2017 book *North Atlantic Right Whales*, opined that the US government "seemed more intent on ensuring no fishing opportunity would be lost than on protecting whales."[20] As the new century began, NMFS continued to focus on gear modification; yet, between 2000 and 2011, 65 new North Atlantic right whale entanglements were documented by observers on planes, boats, and beaches. Entanglement was responsible for 25 percent of right whale deaths during that time. The litany of point and counterpoint, with the conservation lobby overbalanced by the fishing industry looking to maintain the status quo, largely supported by the government, is fully documented in Laist's book. Today, it is time to deal with the reality that entanglement risk isn't about how a line is rigged to retrieve a trap; it is about the presence of the line itself.

The good news is that retrieving traps does not require rope. Technology to bring up hardware from the ocean floor, without the need for a persistent rope from the device to the surface obstructing the water column for days or longer, to be used only when it is time for retrieval, exists and has been used for decades. Since the 1960s, scientists have triggered the recovery of instruments, such as sound and current recorders, from the seafloor by using an acoustic signal to initiate those instruments' ascent. A similar system could be used for trap retrieval. Endlines could be stowed on the seafloor and brought to the surface only for trap retrieval, or the system could trigger the ascent of the trap itself. This type of system has been called on-demand, buoyless, rope-less, or pop-up retrieval. In 1979, Jon Lien described this concept to me as a way of recovering gill nets without an endline. Thirty years later, in 2009, the US Atlantic Large Whale Take Reduction Team, which comprises fishing industry, fisheries management, nongovernmental, and science stakeholders, made a uniquely consensual recommendation, that experimental on-demand fishing should be tried in a closed area of the Great South Channel, east of Nantucket. But NMFS was unable to plan adequate management of the experiment, so it was shelved. A major barrier to the adoption of on-demand fishing has been the cost of acoustic release and trap location technology. Not only does the trap need to be retrieved acoustically, but the role of the surface buoy as a location marker also has to be fulfilled acoustically. Furthermore, before this technology can be used for trap retrieval, long-established laws that require a surface marker wherever traps are set on the bottom need to be revised. Additionally, the residual entanglement risk posed by premature release

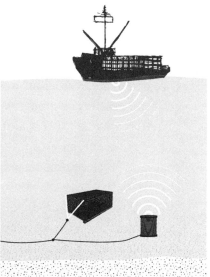

Sketch of the concept of on-demand, ropeless, buoyless, or pop-up fishing. *Left*: Traditional trap fishing requires a line from the trap to a surface marker buoy. This line enables marking of the trap location and owner as well as retrieval of the trap. It also presents a substantial entanglement risk to North Atlantic right whales. *Right*: An on-demand system replaces the vertical line and buoy with an acoustic marker of location and ownership and an acoustically triggered retrieval mechanism. The relatively minor remaining risk of entanglement is with the groundline between traps. Fishing gear not to scale. © Woods Hole Oceanographic Institution, Natalie Renier, WHOI Creative.

of any bottom-stowed line, and by groundline sitting on the bottom, as discussed above, needs to be considered.

Recently, my colleague Mark Baumgartner and I helped to set up a series of workshops to promote on-demand trap retrieval.[21] The challenges for such a technology include making it compatible with a profitable fishery (efficient as well as safe) and able to fulfill the tasks of trap retrieval and location required by all interested stakeholders. One type of system releases a coiled, bagged, or spooled rope from the trap that floats to the surface, allowing the trap to be retrieved in the traditional manner with a hydraulic hauler. Another type of system, which raises the traps by inflation of a lift bag, uniquely avoids the need for any retrieval rope at all. However, the industry is heavily, though not entirely, entrenched in current practices, citing unbearable costs, inefficiency, and fisherman safety as concerns about on-demand technologies. There is a name for this kind of entrenchment: exnovation. The opposite of innovation, exnovation can occur when products and processes shown to be best-in-class are standardized to ensure that they are not innovated further. An example is the widespread denial by oil and gas industries concerning their contribution to climate change when other more sustainable energy options, such as solar, wind, and tidal energy, could be pursued far more aggressively.

While the next proposal is perhaps counterintuitive from the harvest perspective, a reduction of trap fishing effort could benefit both whales and commercial use of the lobster resource. We recently compared yields of lobsters at different levels of fishing effort.[22] For instance, the US lobster fishery in Maine expends approximately 7.5 times as much effort as the Canadian fishery west of Nova Scotia, where harvesters catch about 3.7 times more lobsters per trap. Furthermore, from 2007 to 2013 in Maine, lobster landings doubled as the number of traps fell 10.5 percent, and landings per trap increased by about 125 percent. This change reflects growth in the lobster population, but also suggests that most, if not all, of the lobsters of a size that can be harvested legally are caught, and therefore that the industry is overcapitalized. Massachusetts has achieved record high lobster landings since seasonal closures of trap fishing have been implemented to protect North Atlantic right whales, especially within the areas most affected by the closures.

Thus, economic losses may not accrue with reduced effort. In fact, reduced effort might actually increase profits while enabling conservation of both lobsters and North Atlantic right whales, as it would reduce fuel use, emissions, labor, hardware, and bait cost, as well as bait species decimation. Furthermore, it would reduce the number of on-demand retrieval devices required to reduce rope in the water column. Industry reactions to our study have ranged from absolute dismissal to concurrence.

The missing link is consumers, who do not know enough to demand a seafood product that is caught in a manner that protects whale populations and shields individual whales from the major animal welfare concern that rope entanglement entails, but also maintains viable and sustainable trap fishing industries.

Were we to take the perspective of our grandchildren fifty years in the future, we might respond to our current actions in one of two ways:

1. How could they all have been so shortsighted as to demand affordable lobster and crab rolls, and cheap shipping of goods from overseas, despite the fact that they knew that the North Atlantic right whale species was headed for extinction? All that we have now are the bones of a few skeletons hanging in museums up and down the east coast of North America.

Or

2. Finally, they all saw that there was a way for whales, fisheries, and ships to coexist. It just took some legislative,

regulatory, and political honesty and fortitude, enforcement of existing regulations, government investment, compromise, industrial ingenuity, and consumer education.

As a scientist, I know that it is beyond urgent that we introduce much more widespread measures to mitigate large whale trauma caused by vessels and fishing gear. As a veterinarian, I see the often long, drawn-out trauma to individual right whales caused by ropes from fishing gear as utterly unacceptable. Do we all have the individual and political will to make it right?

Government regulations regarding how many animals can die each year are not being enforced adequately. The balance of political power between the seafood industry and the conservation and animal welfare lobbies is understandably heavily skewed to support the coastal communities that depend on the seafood industry for bringing cash flow to the region. Nevertheless, fishermen have been burdened and fatigued by years of whale conservation–driven mandatory gear modifications, while the problem of whale trauma has continued to get worse.

If the government fails to make its regulations work for either the fishing industry or the whales, what can we, as individuals, do?

Consumers can seek to buy lobsters and crabs that are caught by whale-friendly methods. We need to develop a new paradigm that recognizes truly sustainable fisheries as those that are sustainable both for the harvested species and for other, nontarget species that are impacted by the fishery. Sustainable seafood programs have been largely funded by the seafood industry, and the inevitable conflict of interest leads to surprising certifications of fisheries—those that entangle whales, for instance.

The idea of retrieving traps using on-demand systems, with no persistent rope in the water column, is abhorrent to many in the industry. But if the pain, suffering, and death of whales caused by entanglement were fundamentally unacceptable, and if fishing with on-demand systems were a legal requirement, then fishermen would develop these methods, and the price of their product would reflect the costs.

At a workshop in February 2018, I summarized technological and traditional solutions to the problem of retrieving traps without persistent rope in the water column. Acoustic release technology was being used in a lobster fishery in Australia at the time. But closer to home, there is a time-honored tool that can retrieve traps without an endline. Traditional grapples are heavy hooks, dropped on a line to the ocean floor to catch onto ropes lying on the bottom between traps. They are used widely to retrieve traps for which endlines have been lost, and for routine retrieval of lobster traps in Boston Harbor, where heavy vessel traffic destroys endlines.[23] They are also routinely used in the golden crab fishery in the Gulf Stream east of Florida, in 1,000–2,000 feet (610 m) of water. However, grapples are also used for illegal fishing. Lobstering licenses specify the number of traps and associated surface buoys that a fisherman can use. But some use additional, illegal trap trawls without any endlines and surface buoys, referred to as "sunken trawls," which enable them to set more traps than their license allows. The sunken trawls, which cannot be detected from the surface, are retrieved—hopefully unseen—using a grapple. My talk of grappling at the workshop was met with derision, as it was seen as slow, unsafe, and encouraging an illegal practice. Yet fishing without endlines could be an accepted and respected method. It could be freed of its current criminal stigma with appropriate state and federal laws and regulations. The cost of developing affordable on-demand retrieval technology must be underwritten by government subsidy. To at least fund acoustic marking of traps without endlines would ensure that all interested parties know where traps are deployed despite the lack of surface markers. Then grapples could be used as a familiar, affordable solution if acoustic trap retrieval were cost prohibitive, especially if the acoustic markers were highly accurate.

How far has this concept progressed? Desert Star Systems, EdgeTech, SMELTS, LobsterLift, FioMarine, DBV Technology and Ashored are all acoustically triggered on-demand systems, most of which are commercially available. The majority involve acoustically triggered release of buoyant line stowed on the trap, but others acoustically trigger inflation of a lift bag. As is so often the case, California is leading the charge toward their adoption. Geoff Shester, from the environmental group Oceana, told me in a July 10, 2020, email:

> We've continued to test pop-up (aka ropeless) gear here in the California Dungeness crab fishery since 2018, as progressive fishermen continue to tell me it is promising and realistic. The California Ocean Protection Council just granted $500K to the National Marine Sanctuary Foundation to purchase pop-up systems from several manufacturers to conduct broad-scale testing in the upcoming 2020–21 crab season. I continue to believe pop-up gear is the only viable solution that will allow continued crab harvest in times and places where endangered whales are concentrating here to feed. New proposed California regulations will close areas to traditional crab fishing when whales are present, and will

allow pop-up gear that is shown to be reliable and can be virtually tracked. This new technology has the potential to save endangered whales and sea turtles from extinction while allowing for continued whale-safe fishing not only here but around the world.

In the Gulf of St. Lawrence in 2020, ten snow crab boats used on-demand traps in an otherwise closed area. Results are still under analysis, but appear positive. This is the beginning of on-demand technology that will enable access to areas that would otherwise have to be closed. The technology has great potential to enable the industry to prosper and to minimize trauma to whales at the same time. On Georges Bank, between the Gulf of Maine and the Northwest Atlantic, another trial of on-demand systems is underway. The gear is working, and it's retrievable in heavy seas. On-demand gear is also being tested in Florida and Georgia in the black sea bass and blue crab fisheries, and in Massachusetts, California, Scotland, Ireland, and England for lobster, crab, and prawn fisheries. It's the beginning of a new era.

I am currently raising funds to work with fishermen, government scientists, and colleagues from nongovernmental organizations—all friends—to expand our development of buoyless systems in the local inshore and offshore lobster fisheries. It's a long way from wading around in decomposing dead whales, but it is a natural progression from there. Veterinarians diagnose, treat, and prevent disease. Here, I've written about all three steps in the context of traumatized right whales. In this case, the prevention isn't showing the owner of an overweight dog how to feed it less, or treat its parasites better. For me, it's helping to reduce the risk of trauma to a right whale. But the principle remains the same.

Fishing vessel *Resolve* undertaking trials of buoyless gear, with engineers on deck. © Woods Hole Oceanographic Institution, Craig LaPlante, WHOI Creative.

.   .   .

What have we learned? Progress in right whale conservation is a slow, endless, and incredibly frustrating process of trying to stay ahead of a moving target. The movement of the shipping lanes in the Bay of Fundy in 2003 certainly reduced the collision risk for right whales in the Bay of Fundy, but once the whales moved to the Gulf of St. Lawrence in the next decade, they encountered a whole new series of risks, both from vessel strikes and from entanglement in lobster gear, as well as the much heavier snow crab gear. The consequences between 2017 and 2020 were especially bad, with known losses of thirty-one North Atlantic right whales in the United States and Canada in those four years. In 2020, the combination of adaptive management with reduced vessel traffic and reduced snow crab fishing resulting from the coronavirus pandemic appeared to alleviate the mortality problem in the Gulf of St. Lawrence, but sadly, a right whale killed by a vessel was found outside of New York Harbor in June 2020. It had first been sighted as a newborn calf off Georgia in December 2019, just seven months earlier. Ironically, acoustic sensors were alerting us to the presence of right whales in that general area at that time, but adequate adaptive management changes were obviously not in place. This death reinforces the strong need for widespread mitigation measures for both vessel and entanglement risks. In the Gulf of St. Lawrence, Canada is now (in 2020) using visual sightings and acoustic detection of whales to trigger mandatory dynamic mitigation requirements for the shipping and fishing industries.

Thus, in continental shelf waters from the Gulf of St. Lawrence to Florida, where right whales spend most of their time, there must be widespread vessel speed restrictions and removal of rope from the water column. These measures have to become the norm over months, years, decades, and centuries if the species is to survive, recover, and thrive, as has been shown to be possible in the case of the Southern right whale.

Much of this book has reflected on time frames: the minutes it takes for a ship or harpoon to kill a whale; the months it takes to kill one with entangling rope; the increasing numbers of years it takes for a female right whale to become fat and fit enough to conceive, gestate, and suckle a calf; the decades over which these females could be reproductively active if we stopped killing them as teenagers; the thousand years over which we have been killing North Atlantic right whales, first with harpoons and now with vessel propellers and rope; to the millennia over which the Iñupiat and other Arctic peoples have successfully managed the bowhead whale population as a key piece of their cultures. This last is the time frame we have to consider when deciding what we have to do for the North Atlantic right whale if we, as capitalist consumers, are to become ethical and successful stewards of this world.

Of course, this right whale story is about just one species, but the evident need for respect, restraint, understanding, and preventive action has to be heard universally. It applies to climate change, fossil fuel impacts, plastic pollution, and human poverty, disease, racial inequality, and hunger—and so much more.

Why are we failing the North Atlantic right whale species— and failing in so many other ways—so badly? We as a species are shortsighted and self-centered. We are focused on instant gratification. We think we need all those goods that come in large ships across the ocean, fuel for our cars and houses, luxury seafood for our restaurants.

I write this six months into the 2020 coronavirus pandemic, during which much of the world stood still before parts of it, including much of the United States, re-erupted into a fresh wave of misguided, selfish self-gratification. Failing to the take the long view. Refusing to recognize that prophylactic measures such as face masks were a necessary short-term cost for long-term gain. In just the same way, if we are to avoid losing the race to save the North Atlantic right whale species, we must take the long view. Following the example of the Arctic peoples, we must show restraint and respect for our ocean resources, such as the North Atlantic right whale. The whale is an icon, just as the lobster and its harvesters are icons. They can all thrive. In learning those attitudes, we can coexist sustainably with the natural world on our fragile planet.

What if we lose the North Atlantic right whale species? It would be another nail in the coffin of the human species. If we continue to destroy biodiversity, what kind of a world will be left for us to survive in? It is an ethical, moral, and practical necessity that humans stop destroying and overexploiting the diverse resources of the globe. If we do not, we will lose a species that is an icon, just like the lobster, snow crab, and cod. Furthermore, whales are an important ocean resource,: they recycle nutrients from the seafloor, fertilizing the food web that fish feed on, delight ecotourists on whale-watching cruises, and provide us with other services that we have no idea of. Their loss would doubtless have unintended consequences of which we are unaware. It's up to all of us to decide if we care about sustaining a diverse and healthy planet.

But above all, within the narrow walls of this book, we have to recognize that the pain and drawn-out suffering that rope entanglement causes to these animals is absolutely unacceptable, and is the ultimate reason why we have to stop what is an utter nightmare for individual, chronically entangled animals.

Postscript 1: Getting Really Cold

Right whale #2301: Born 1993, had one calf in 2002, died
entangled in rope, 2005.

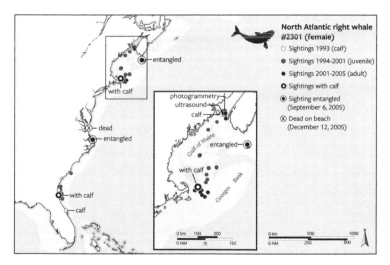

Track of North Atlantic right whale #2301 from when she was seen as a calf off
Florida, to entanglement in Canada, to when she died and washed ashore in
Virginia. Data: North Atlantic Right Whale Consortium; plot: © Woods Hole
Oceanographic Institution, Natalie Renier, WHOI Creative.

Timeline series of photographs of #2301. *Top left*: As a calf with mother. Lindsay Hall, New England Aquarium. *Top right*: Aerial view. Heather Pettis, NEFSC. *Bottom left*: View of right side of head. Jessica Damon, New England Aquarium. *Bottom right*: Entangled in the Bay of Fundy. New England Aquarium. Permit: NMFS #775-1600-2.

Spring 1993: *Life as a young right whale calf consists of sticking close to mom and doing what she says. Sometimes when mom is feeding deeper than I want to dive, I stay nearer the surface till she comes back up. Suckling is really important to get at all that oil-rich milk.*

1994–2002: *I hung out with my mother for about a year. It took a while to learn how to feed on my own once I wasn't getting so much milk, so I lost some weight after I was weaned. But I gradually got the idea of how to swim through the water with my mouth wide open. It's really tiring to swim like that. We have to swim very hard to make progress through the water, and we have to be very picky about where we feed, so that we can lay away more energy than we expend in catching the food. Learning to find the best places to eat was tough, but I found that I have an innate sense of where to go, how to find places that we know have been bountiful in years past. I grew in length and width, and hung out with other right whales in large groups. One day I realized I was pregnant, so it was time to head south to warmer waters. My first calf was born, and it did well.*

2004: *Once my calf was weaned, I went back to work, building up my body condition, hoping to get fat enough to have another calf soon, if I found good food, or some years later if I didn't. One of the things my mother had told me, time and time again, was to be wary of ships, boats, and fishing gear. The ships were very noisy and confusing, because we could hear them for hours and hours as they got closer, and we got to the point where we simply ignored them. Then usually as they went by, the noise slowly faded away again. But sometimes, as it got louder and louder, the noise dropped off abruptly, which was a great, but deceptive, relief. My mother had explained that it meant that the bow of the ship was shielding the noise from the propeller and engine, which were at the back of the ship. Her advice was to dive as fast as possible to avoid being hit. My mother also warned me about rope. Fishing boats had all kinds of different ropes and nets. The ones that dragged nets weren't a major threat to us, but they did a number on the fish-eating whales and dolphins. The biggest issue for us was rope that went from the surface of the ocean to the bottom, where it was attached to one or more wire mesh cages containing crabs and lobsters. We also had to watch out for long curtains of net, stretched along the bottom, with weights to keep them in place, and floats to keep the net standing up in the water. When I was a calf, I used to try and play with ropes, and I would get a serious telling off from my mother. But once I was weaned, I found it really hard to miss the ropes when I was so intent on finding the best food patch and sticking with it, twisting and turning all the while. Often, I would brush against a rope running from the bottom to the surface. Usually it would just glide by, but sometimes I would get hooked up in it. I learned to stop, and back off, although the temptation to twist and turn and thrash was sometimes irresistible. I picked up a few minor scars.*

When it happened, my calf had been weaned for less than a year, so I was still skinny and hungry all the time. I had found a really good patch of food. I can't say where I was. I ran into some fairly light line, but before I knew it, I was tangled up. I had it in my mouth, over my blowhole, and tightly wrapped around my left flipper. I'm ashamed to say that I panicked. I twisted, turned, rolled, thrashed, and charged around like a crazy whale. I was able to get the trap hooked on a rock and broke that off, but I was still left with an ugly mess of rope. I could feel all the knots and twists. Every time I swam, I felt fresh stabbing pain, especially around my left flipper. I could see the rope cutting into the flesh, and eventually, I felt it start to cut into the bone. My left blowhole was becoming increasingly useless. I carried that rope around for weeks and months.

January 2005: *My left flipper felt like it was dead. Lice had spread out around my blowhole and left flipper. I was losing all that beautiful fat I had been laying down for my next calf. I felt desperate. The last thing I remember was heading south to warmer water, as I was getting really cold with no fat between the water and myself.*

Postscript 2: A Lonely Tunnel
with No Light at the End

Right whale #2030: Born before 1990, no calves, died 1999. 189

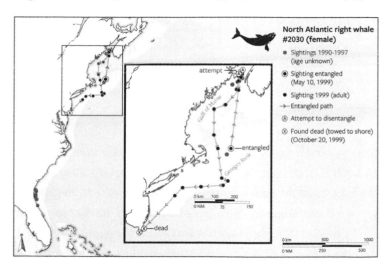

Track of North Atlantic right whale #2030 from when she was first sighted, to entanglement off Nantucket, to when she died east of Cape May, New Jersey. Data: Center for Coastal Studies; plot: © Woods Hole Oceanographic Institution, Natalie Renier, WHOI Creative.

Timeline series of photographs of #2030. *Top left*: Roseway Basin, Nova Scotia, September 1991. Chris Slay, New England Aquarium. *Top right*: Bay of Fundy, September 1996. Philip Hamilton, New England Aquarium. *Bottom left*: Entangled east of Nantucket, May 1999. Northeast Fisheries Science Center. *Bottom right*: Disentanglement attempt, September 4, 1999. Chris Slay, New England Aquarium. Permit: NMFS #775-1600-2.

June 1990: *I'd found a large patch of surface plankton in Massachusetts Bay, so I was skimming along, with my upper jaw above the water. I could see a boat coming. The propeller from the boat scattered the plankton patch, so I had to work harder to get the food I needed. But the plankton was good, the winds quite light, and the water warming up, so I stuck around that general area for about ten days. Then I was approached by the same boat and a few others as well.*

September 1991: *Food in the Bay of Fundy was all really deep where I was swimming, so the boats didn't mess with it. There, the challenge was to get down to it. It was often near or on the bottom, under 600 feet of water. The tide there is really strong, but if you find the right places, it pushes all the food into really tight balls and streaks. This can make it a stream of pure oil. It's the best! Sometimes you have to get on your side, or upside down, to get to the thickest part, and stick your upper jaw into the mud. Open your mouth, and thick plankton streams pile into your baleen hairs, and all it takes is a regular swish of the tongue and down it all goes. But at that depth you get woozy pretty quick, so you have to spend quite a bit of time back up at the surface to recover your breath.*

August 1995: *I stayed in the Bay of Fundy through September of that year. A green boat watched us all as we gathered in big groups at the surface, where we hang out and find mates. This year, I was old enough and fat enough to take part. Indeed, by the end of the month, I was getting all the attention. It was time to get pregnant and have a calf.*

May–October 1999: *One dark night I was busy feeding, and bang, I swam full tilt into a gill net. I managed to get my head under it, but I got the headrope over my back. I then made the mistake of spinning. My mother had always said not to do that on any count, but the net was just so annoying, and I thought I could break out of it. But it was not to be. I got net and rope tightly wrapped around both of my flippers, with the net spread across my back behind my blowholes. I was able to breathe, so that was good, but it was hard to swim and almost impossible to feed.*

I knew right away that I was in trouble, but what surprised me was the pain. Most times when I ran my head into the bottom, or had a collision with another whale, the pain had gone away pretty quick, but this was different. It just kept on jabbing, and as I swam through the next weeks and months, it got worse and worse. Each day as I swam and tried to feed, the rope tightened around my flippers and dug deeper and deeper into my back. It was hard to expand my chest fully, which made the quick breaths we take, to minimize the amount of seawater we inhale, especially hard. My lungs started to burn as I began to feel heavy in my chest. And having ever-tightening wraps of rope and net around both armpits that were linked over my back made it almost impossible to move either or both of my flippers adequately when I needed to change my tilt, roll, or heading. Basically, it felt like I was swimming while constrained in a tube. It was so hard to do what I needed to do to feed and move between areas where the food was good. I lost contact with the other whales that I usually spent time with. I was in a lonely tunnel with no light at the end.

For about ten days I was persistently approached by a bunch of different boats. One had two long poles sticking out of its bows, and as I surfaced, it tried to come up behind me and drop a loop of rope over my tail flukes. I didn't understand what they were trying to do, but the last thing I needed was more rope wrapping me up, so I did everything I could to foil their efforts, and was largely successful. Then they tied large objects onto the small amount of rope that was trailing from my body. Those things made it really hard to dive, and I was already really, really tired from the months of living with the rope as it slowly ate into my back and armpits. Maybe those humans were trying to help, but the whole thing made my blood boil. I did everything I could to rid myself of the foul rope and net that I had blundered into in the first place, and to avoid the humans who seemed hell-bent on adding more rope.

It was time to go find some peace and quiet, so I wandered south. Over the next five weeks I bumbled around, slowly getting weaker and weaker. As the water temperature dropped, I started to get cold. My blubber coat had wasted away. The rope continued to cut away the blubber over my shoulder blades. I was in a delirious state on a shoal when another boat came along.

Acknowledgments

This book was made possible by Hannah. It is dedicated to all the whales that could have been, and will be, and to my lab cohort: Carolyn Miller, Regina Campbell-Malone, Nadine Lysiak, Andrea Bogomolni, and Julie van der Hoop. It also recalls the work of hundreds of other colleagues. Joe Calamia, my editor at the University of Chicago Press, sought me out and showed me what I was trying to say. Norma Sims Roche rescued and polished my twisted prose. I thank three anonymous reviewers for their constructive comments. The book was funded by the Volgenau Foundation and a Woods Hole Oceanographic Institution Stanley Watson Chair. Funding to the author for research described herein came from institutions including International Fund for Animal Welfare, North Pond Foundation, M. S. Worthington Foundation, Herrington-Fitch Family Foundation, Annenberg Foundation, Northeast Consortium, National Oceanic and Atmospheric Administration, US Office of Naval Research, International Whaling Commission, the Massachusetts Environmental Trust, and the SeaWorld Busch Gardens Conservation Fund. Use of North Atlantic right whale data and images was approved by the North Atlantic Right Whale Consortium.[1] Research was permitted by the United States and Canadian governments as detailed in figure captions.

Notes

1   S. M. Sharp, W. A. McLellan, D. S. Rotstein, A. M. Costidis, S. G.
    Barco, K. Durham, T. D. Pitchford et al., "Gross and Histopatho-
    logic Diagnoses from North Atlantic Right Whale *Eubalaena
    glacialis* Mortalities between 2003 and 2018," *Diseases of Aquatic
    Organisms* 135, no. 1 (2019): 1–31, https://www.int-res.com
    /abstracts/dao/v135/n1/p1-31/.

2   M. J. Crone and S. D. Kraus, *Right Whales* (Eubalaena glacialis),
    in the Western North Atlantic: A Catalog of Identified Individuals
    (Boston: North Atlantic Right Whale Consortium, New England
    Aquarium, 1990), 225, http://rwcatalog.neaq.org.

3   R. Reeves, T. Smith, and E. Josephson, "Near-Annihilation of
    a Species: Right Whaling in the North Atlantic," in *The Urban
    Whale: North Atlantic Right Whale at the Crossroads*, ed. S. Kraus
    and R. Rolland (Cambridge, MA: Harvard University Press,
    2007), 39–74.

4   C. R. Markham, "On the Whale-Fishery of the Basque Provinces
    of Spain," *Proceedings of the Zoological Society of London* (1881):
    969–76, https://doi.org/10.1111/j.1096-3642.1881.tb01354.x.

5   G. E. Maul and D. E. Sergeant, "New Cetacean Records from
    Madeira," *Bocagiana* 43 (1977): 1–8, https://www.researchgate
    .net/publication/257304039_New_records_of_cetacean_species
    _for_Madeira_Archipelago_with_an_updated_checklist.

6   H. Pettis, R. Pace, and P. Hamilton, North Atlantic Right Whale
    Consortium 2020 Annual Report Card, https://www.narwc.org
    /report-cards.html.

CHAPTER ONE

1 M. J. Moore, S. Landry, B. Bowman, A. R. Knowlton, P. K. Hamilton, and D. S. Rotstein, "Morbidity and Mortality of Chronically
   Entangled North Atlantic Right Whales: A Major Welfare Issue"
   (16th Biennial Conference on the Biology of Marine Mammals,
   Society for Marine Mammals, San Diego, CA, 2005), https://
   www.marinemammalscience.org/wp-content/uploads/2014
   /09/Abstracts-SMM-Biennial-San-Diego-2005.pdf.

CHAPTER TWO

1 "Development of Cod Traps," History of the Northern Cod
   Fishery, https://www.cdli.ca/cod/history7.htm.
2 "Regina Maris," Wikipedia, https://en.wikipedia.org/wiki/Regina
   Maris(schooner).
3 S. Katona, B. Baxter, O. Brazier, S. Kraus, J. Perkins, and H. Whitehead, "Identification of Humpback Whales by Fluke Photographs," in Behavior of Marine Animals, ed. H. E. Winn and B. L.
   Olla (Boston: Springer, 1979), 33–44.
4 Examples include Happywhale, https://happywhale.com/, and
   Flukebook, https://www.flukebook.org/.
5 H. Whitehead, Sperm Whales: Social Evolution in the Ocean (Chicago: University of Chicago Press, 2003).
6 S. Kemp and A. G. Bennett, "On the Distribution and Movements
   of Whales on the South Georgia and South Shetland Whaling
   Grounds," Discovery Reports 6 (1932): 165–90.
7 D. G. Burnett, The Sounding of the Whale: Science and Cetaceans
   in the Twentieth Century (Chicago: University of Chicago Press,
   2012).
8 "Við Áir," Wikipedia, https://en.wikipedia.org/wiki/Vi%C3%B0
   _%C3%81ir.
9 M. J. Moore, "How We All Kill Whales," ICES Journal of Marine
   Science: Journal du Conseil 71 (2014): 760–63, https://doi.org/10
   .1093/icesjms/fsu008.
10 R. Payne, O. Brazier, E. M. Dorsey, J. S. Perkins, V. J. Rowntree,
   and A. Titus, "External Features in Southern Right Whales (Eubalaena australis) and Their Use in Identifying Individuals," in

Communication and Behavior of Whales, ed. R. Payne, US Marine Mammal Commission (Boulder, CO: Westview Press, 1983), 643.

11   "Baccalieu Island," Wikipedia, https://en.wikipedia.org/wiki /Baccalieu_Island.

12   "Capelin," Fisheries and Oceans Canada, http://www.dfo-mpo .gc.ca/species-especes/profiles-profils/capelin-capelan-eng .html.

13   L. Pendergast, "The Whale Man: The Exceptional Legacy of Jon Lien," *Gazette*, Memorial University, April 11, 2019, https:// gazette.mun.ca/campus-and-community/the-whale-man/.

14   "Whale Entanglement: Building a Global Response," International Whaling Commission, https://iwc.int/entanglement.

15   A trailer for the play provides important insights into Jon Lien: "Between Breaths," *Here & Now*, https://www.youtube.com /watch?v=qinXwaMqu3w.

16   P. Tyack and H. Whitehead, "Male Competition in Large Groups of Wintering Humpback Whales," *Behavior* 83 (1983): 132–54, https://doi.org/10.1163/156853982X00067.

17   H. Whitehead and M. J. Moore, "Distribution and Movements of West Indian Humpback Whales in Winter," *Canadian Journal of Zoology* 60, no. 9 (1982): 2203–11.

CHAPTER THREE

1   K. Parry, M. Moore, and C. G. Hulland, "Why Do Whales Come Ashore?," *New Scientist*, March 17, 1983, 716–17.

2   "History and Purpose," International Whaling Commission, https://iwc.int/history-and-purpose.

3   R. Lambertsen and M. Moore, *Behavioral and Post Mortem Observations on Fin Whales Killed with Explosive Harpoons with Preliminary Conclusions concerning Killing Efficiency* (International Whaling Commission Technical Report, TC/36/HK, 1983), 1–23.

4   J.-P. Desforges, A. Hall, B. McConnell, A. Rosing-Asvid, J. L. Barber, A. Brownlow, S. De Guise et al., "Predicting Global Killer Whale Population Collapse from PCB Pollution," *Science* 361, no. 6409 (2018): 1373–76.

5   J. Bockstoce, *Whales, Ice and Men: The History of Whaling in the Western Arctic* (Seattle: University of Washington Press, 1995).

6 B. Demuth, *Floating Coast: An Environmental History of the Bering Coast* (New York: W. W. Norton, 2019).

CHAPTER FOUR

1 J. Roman, J. A. Estes, L. Morissette, C. Smith, D. Costa, J. McCarthy, J. Nation et al., "Whales as Marine Ecosystem Engineers," *Frontiers in Ecology and the Environment* 12, no. 7 (2014): 377–85, https://esajournals.onlinelibrary.wiley.com/doi/10.1890/130220.

2 "Wild Whale Spins on Command of Diver," YouTube, https://youtu.be/EPtrYuexRjk.

3 NARWC Annual Report Cards, North Atlantic Right Whale Consortium, https://www.narwc.org/report-cards.html.

4 D. G. Burnett, *The Sounding of the Whale: Science and Cetaceans in the Twentieth Century* (Chicago: University of Chicago Press, 2012).

5 J. Gabbatiss, "Iceland Announces Plan to Kill over 2,000 Whales within Next Five Years," *Independent*, February 21, 2019, https://www.independent.co.uk/environment/whaling-iceland-whales-kill-fin-minke-new-quota-hunting-japan-a8790491.html.

6 T. Adalbjornsson, "Meet Iceland's Whaling Magnate," *New York Times*, August 10, 2018, https://www.nytimes.com/2018/08/10/climate/iceland-whaling.html.

7 "Commercial Whaling in Newfoundland and Labrador in the 20th Century," Heritage: Newfoundland & Labrador, https://www.heritage.nf.ca/articles/environment/whaling-in-the-20th-century.php.

8 "Whaling Stations," Government of South Georgia & the South Sandwich Islands, http://www.gov.gs/heritage-2/whaling-stations/.

9 S. W. Stoker and I. I. Krupnik, "Subsistence Whaling," in *The Bowhead Whale*, ed. J. Burns, J. J. Montague, and C. Cowles (Lawrence, KS: Society for Marine Mammalogy, 1993), 579–629.

10 Stoker and Krupnik, "Subsistence Whaling."

11 C. George, personal communication, December 23, 2019.

12 Stoker and Krupnik, "Subsistence Whaling."

13 J. Burns, "Epilogue," in *The Bowhead Whale*, ed. Burns, Montague, and Cowles, 745–62.

14   G. Ross, "Commercial Whaling in the North Atlantic Sector," in *The Bowhead Whale*, ed. Burns, Montague, and Cowles, 511–61.

15   A. Aguilar, "A Review of Old Basque Whaling and Its Effect on the Right Whales (*Eubalaena glacialis*) of the North Atlantic," *Reports of the International Whaling Commission* 10 (special issue) (1986): 191–200.

16   C. R. Markham, "On the Whale-Fishery of the Basque Provinces of Spain," *Proceedings of the Zoological Society of London* (1881): 969–76, https://doi.org/10.1111/j.1096-3642.1881.tb01354.x.

17   Ross, "Commercial Whaling in the North Atlantic Sector."

18   J. R. Bockstoce and J. J. Burns, "Commercial Whaling in the North Pacific Sector," in *The Bowhead Whale*, ed. Burns, Montague, and Cowles, 563–77.

19   G. Givens, S. Edmondson, J. George, R. Suydam, R. Charif, A. Rahaman, D. Hawthorne et al., "Estimate of 2011 Abundance of the Bering-Chukchi-Beaufort Seas Bowhead Whale Population" (Paper SC/65a/BRG01, Scientific Committee of the International Whaling Commission 65a, Jeju Island, Korea, 2013).

20   W. Streever, *Adventures in the World's Frozen Places* (New York: Little, Brown, 2009); Eli Kintisch, "History Is Melting," *Hakai*, January 26, 2016, https://www.hakaimagazine.com/features/history-melting/.

21   B. A. McLeod, M. W. Brown, M. J. Moore, W. Stevens, S. H. Barkham, M. Barkham, and B. N. White, "Bowhead Whales, and Not Right Whales, Were the Primary Target of 16th- to 17th-Century Basque Whalers in the Western North Atlantic," *Arctic* (2008): 61–75, https://www.jstor.org/stable/40513182.

22   Northern Alaska Sea Ice Project Jukebox, University of Alaska Fairbanks Oral History Program, https://jukebox.uaf.edu/site7/seaice.

23   M. L. Druckenmiller, H. Eicken, J. C. George, and L. Brower, "Assessing the Shorefast Ice: Iñupiat Whaling Trails off Barrow, Alaska," in *SIKU: Knowing Our Ice: Documenting Inuit Sea Ice Knowledge and Use*, ed. I. Krupnik et al. (Dordrecht: Springer Netherlands, 2010), 203–28.

24   K. Brewster, interview with C. George, part 1, June 4, 2017, Northern Alaska Sea Ice Project Jukebox, https://jukebox.uaf.edu/site7/interviews/2993; interview with C. George, part 2, June 4, 2017, https://jukebox.uaf.edu/site7/interviews/2994.

25   Brewster, interview with George, part 2, June 4, 2017.

26 Brewster, interview with George, part 2, June 4, 2017.

27 K. Brewster, interview with L. Brower, February 26, 2016, Northern Alaska Sea Ice Project Jukebox, https://jukebox.uaf.edu /site7/interviews/2782.

28 K. Brewster and O. Dammann, interview with W. Adams, November 12, 2013, Northern Alaska Sea Ice Project Jukebox, https://jukebox.uaf.edu/site7/interviews/2061.

29 K. Brewster, interview with J. Adams, interview 2, part 1, November 15, 2013, Northern Alaska Sea Ice Project Jukebox, https:// jukebox.uaf.edu/site7/interviews/2099.

30 "Whaling Voyages," Iñupiat Heritage Center, April 14, 2015, https://www.nps.gov/inup/details.htm.

31 North Atlantic Right Whale Consortium, https://www.narwc .org/.

32 Alaska Eskimo Whaling Commission, "Ilitqusia Aġviġum: Spirit of the Whale; a Way of Life for the Iñupiat and Yupik People," 2012, http://nebula.wsimg.com/9473dd5c3a0d8838e146bdd2a5 5d7d86?AccessKeyId=5111461408D77964347D&disposition=0 &alloworigin=1.

CHAPTER FIVE

1 M. C. I. Martins, L. Sette, E. Josephson, A. Bogomolni, K. Rose, S. M. Sharp, M. Niemeyer, and M. Moore, "Unoccupied Aerial System Assessment of Entanglement in Northwest Atlantic Gray Seals (*Halichoerus grypus*)," *Marine Mammal Science* 35 (2019): 1613–24.

2 "U.S. Route 6," Wikipedia, https://en.wikipedia.org/wiki/U.S. _Route_6.

3 M. J. Moore, J. van der Hoop, S. G. Barco, A. M. Costidis, F. M. Gulland, P. D. Jepson, K. T. Moore et al., "Criteria and Case Definitions for Serious Injury and Death of Pinnipeds and Cetaceans Caused by Anthropogenic Trauma," *Diseases of Aquatic Organisms* 103, no. 3 (2013): 229–64, https://doi.org/10.3354/dao02566.

4 W. A. McLellan, S. A. Rommel, M. J. Moore, and D. A. Pabst, *Right Whale Necropsy Protocol* (final report to NOAA Fisheries for contract #40AANF112525, US Department of Commerce, National Oceanic and Atmospheric Administration, 2004), 51

pp., available from NOAA Fisheries Service, http://citeseerx.ist
.psu.edu/viewdoc/download?doi=10.1.1.296.8760&rep=rep1&
type=pdf.

5   M. J. Moore, "How We All Kill Whales," *ICES Journal of Marine
Science: Journal du Conseil* 71 (2014): 760–63, https://doi.org/10
.1093/icesjms/fsu008.

CHAPTER SIX

1   M. Moore (@MklJMoore), Twitter, August 20, 2020, 9:00 p.m.,
    https://twitter.com/MklJMoore/status/1296613103694761985.
2   D. Mattila, personal communication, September 15, 2019.
3   North Atlantic Right Whale Consortium, https://www.narwc
    .org/.
4   M. Moore, A. Knowlton, S. Kraus, W. McLellan, and R. Bonde,
    "Morphometry, Gross Morphology and Available Histopathology
    in Northwest Atlantic Right Whale (*Eubalaena glacialis*) Mortali-
    ties (1970 to 2002)," *Journal of Cetacean Research and Manage-
    ment* 6 (2004): 199–214.
5   M. J. Moore, A. Stamper, S. D. Kraus, and R. Rolland, eds.,
    "Report of a Workshop on Large Whale Medical Intervention:
    Indications and Technology Development," working paper, Feb-
    ruary 29, 2000, http://hdl.handle.net/1912/2997.
6   W. E. Schevill, C. Ray, K. W. Kenyon, R. T. Orr, and R. G. Van
    Gelder, "Immobilizing Drugs Lethal to Swimming Mammals,"
    Science 157, no. 3789 (1967): 630–31.
7   M. Moore, M. Walsh, J. Bailey, D. Brunson, F. Gulland, S. Landry,
    D. Mattila et al., "Sedation at Sea of Entangled North Atlantic
    Right Whales (*Eubalaena glacialis*) to Enhance Disentangle-
    ment," *PLoS ONE* 5, no. 3 (2010): e9597, https://doi.org/10.1371
    /journal.pone.0009597.
8   Schevill, Ray, Kenyon, Orr, and Van Gelder, "Immobilizing Drugs
    Lethal to Swimming Mammals."
9   Moore, Walsh, Bailey, Brunson, Gulland, Landry, Mattila et al.,
    "Sedation at Sea of Entangled North Atlantic Right Whales."
10 D. Mattila, personal communication, September 15, 2019.
11 Moore, Walsh, Bailey, Brunson, Gulland, Landry, Mattila et al.,
    "Sedation at Sea of Entangled North Atlantic Right Whales."

12   Moore, Walsh, Bailey, Brunson, Gulland, Landry, Mattila et al., "Sedation at Sea of Entangled North Atlantic Right Whales."

13   M. J. Moore, G. H. Mitchell, T. K. Rowles, and G. Early, "Dead Cetacean? Beach, Bloat, Float, Sink," *Frontiers in Marine Science* 7, no. 333 (2020), https://doi.org/10.3389/fmars.2020.00333.

14   Moore, A. Knowlton, Kraus, McLellan, and Bonde, "Morphometry, Gross Morphology and Available Histopathology in Northwest Atlantic Right Whale (*Eubalaena glacialis*) Mortalities (1970 to 2002)."

15   M. J. Moore, D. Reeb, C. Miller, and D. S. Smith, "Large Whale Disentanglement Technology Workshop," 2001, http://hdl .handle.net/1912/2998.

16   F. M. D. Gulland, F. Nutter, K. Dixon, J. Calambokidis, G. Schorr, J. Barlow, T. Rowles et al., "Health Assessment, Antibiotic Treatment, and Behavioral Responses to Herding Efforts of a Cow-Calf Pair of Humpback Whales (*Megaptera novaeangliae*) in the Sacramento River Delta, California," *Aquatic Mammals* 34 (2008): 182-92.

17   M. J. Moore, How we can all stop killing whales: a proposal to avoid whale entanglement in fishing gear. *ICES Journal of Marine Science* 76 (2019): 781-86, fig. 4c, https://doi.org/10.1093/ices jms/fsy194.

CHAPTER SEVEN

1   W. A. McLellan, S. A. Rommel, M. J. Moore, and D. A. Pabst, *Right Whale Necropsy Protocol* (final report to NOAA Fisheries for contract #40AANF112525, US Department of Commerce, National Oceanic and Atmospheric Administration, 2004), 51 pp., available from NOAA Fisheries Service, http://citeseerx.ist .psu.edu/viewdoc/download?doi=10.1.1.296.8760&rep=rep1& type=pdf.

2   S. M. Sharp, W. A. McLellan, D. S. Rotstein, A. M. Costidis, S. G. Barco, K. Durham, T. D. Pitchford et al., "Gross and Histopathologic Diagnoses from North Atlantic Right Whale *Eubalaena glacialis* Mortalities between 2003 and 2018," *Diseases of Aquatic Organisms* 135, no. 1 (2019): 1-31, https://doi.org/10.3354/dao 03376.

3   M. Moore, R. Andrews, T. Austin, J. Bailey, A. Costidis, C. George, K. Jackson et al., "Rope Trauma, Sedation, Disentanglement, and Monitoring-Tag Associated Lesions in a Terminally Entangled North Atlantic Right Whale (*Eubalaena glacialis*)," *Marine Mammal Science* 29, no. 2 (2012): e98–e113, http://dx.doi .org/10.1111/j.1748-7692.2012.00591.x.

4   M. J. Moore and A. N. Zerbini, "Dolphin Blubber/Axial Muscle Shear: Implications for Rigid Transdermal Intramuscular Track- ing Tag Trauma in Whales," *Journal of Experimental Biology* 220, no. 20 (2017): 3717–23.

5   M. Johnson and P. Tyack, "A Digital Acoustic Recording Tag for Measuring the Response of Wild Marine Mammals to Sound," *IEEE Journal of Oceanic Engineering* 28 (2003): 3–12, doi: 10.1109 /JOE.2002.808212.

6   J. van der Hoop, M. Moore, A. Fahlman, A. Bocconcelli, C. George, K. Jackson, C. Miller et al., "Behavioral Impacts of Disentanglement of a Right Whale under Sedation and the Ener- getic Cost of Entanglement," *Marine Mammal Science* 30, no. 1 (2014): 282–307, https://doi.org/10.1111/mms.12042.

7   van der Hoop, Moore, Fahlman, Bocconcelli, George, Jackson, Miller et al., "Behavioral Impacts of Disentanglement of a Right Whale under Sedation and the Energetic Cost of Entanglement."

8   J. M. van der Hoop, P. Corkeron, J. Kenney, S. Landry, D. Morin, J. Smith, and M. J. Moore, "Drag from Fishing Gear Entangling North Atlantic Right Whales," *Marine Mammal Science* 32, no. 2 (2016): 619–42, https://doi.org/10.1111/mms.12292.

9   J. M. van der Hoop, D. P. Nowacek, M. J. Moore, and M. Trian- tafyllou, "Swimming Kinematics and Efficiency of Entangled North Atlantic Right Whales," *Endangered Species Research* 32 (2017): 1–17, https://doi.org/10.3354/esr00781.

10 J. van der Hoop, P. Corkeron, and M. Moore, "Entanglement Is a Costly Life-History Stage in Large Whales," *Ecology and Evolution* 7, no. 1 (2017): 92–106, https://doi.org/10.1002/ece3.2615.

11 A. Knowlton, P. Hamilton, M. Marx, H. Pettis, and S. Kraus, "Monitoring North Atlantic Right Whale *Eubalaena glacialis* Entanglement Rates: A 30 Yr Retrospective," *Marine Ecology Progress Series* 466 (2012): 293–302, https://doi.org/10.3354 /meps09923.

CHAPTER EIGHT

1   S. Monsarrat, M. G. Pennino, T. D. Smith, R. R. Reeves, C. N.
     Meynard, D. M. Kaplan, and A. S. L. Rodrigues, "A Spatially
     Explicit Estimate of the Prewhaling Abundance of the Endan-
     gered North Atlantic Right Whale," *Conservation Biology* 30,
     no. 4 (2016): 783–91, https://doi.org/10.1111/cobi.12664.

2   H. Pettis, R. Pace, and P. Hamilton, North Atlantic Right Whale
     Consortium 2020 Annual Report Card, https://www.narwc.org
     /report-cards.html; R. M. Pace, P. J. Corkeron, and S. D. Kraus,
     "State–Space Mark–Recapture Estimates Reveal a Recent
     Decline in Abundance of North Atlantic Right Whales," *Ecology
     and Evolution* (2017): 8730–41, https://doi.org/10.1002/ece3
     .3406.

3   D. G. Burnett, *The Sounding of the Whale: Science and Cetaceans
     in the Twentieth Century* (Chicago: University of Chicago Press,
     2012).

4   "Introduction to IMO," International Maritime Organization,
     https://www.imo.org/.

5   A. S. M. Vanderlaan, J. Corbett, S. Green, J. Callahan, C. Wang,
     R. Kenney, C. Taggart, and J. Firestone, "Probability and Miti-
     gation of Vessel Encounters with North Atlantic Right Whales,"
     Endangered Species Research 6 (2009): 273–85, https://doi.org/10
     .3354/esr00176.

6   J. M. van der Hoop, A. S. M. Vanderlaan, and C. T. Taggart, "Abso-
     lute Probability Estimates of Lethal Vessel Strikes to North Atlan-
     tic Right Whales in Roseway Basin, Scotian Shelf," *Ecological
     Applications* 22, no. 7 (2012): 2021–33, https://doi.org/10.1890
     /11-1841.1.

7   A. S. M. Vanderlaan and C. T. Taggart, "Efficacy of a Voluntary
     Area to Be Avoided to Reduce Risk of Lethal Vessel Strikes to
     Endangered Whales," *Conservation Biology* 23, no. 6 (2009):
     1467–74, https://doi.org/10.1111/j.1523-1739.2009.01329.x.

8   N. Record, J. Runge, D. Pendleton, W. Balch, K. Davies, A. Persh-
     ing, C. Johnson et al., "Rapid Climate-Driven Circulation
     Changes Threaten Conservation of Endangered North Atlantic
     Right Whales," *Oceanography* 32, no. 2 (2019): 162–69; K. T.
     Davies, M. W. Brown, P. K. Hamilton, A. R. Knowlton, C. T. Tag-
     gart, and A. S. Vanderlaan, "Variation in North Atlantic Right
     Whale *Eubalaena glacialis* Occurrence in the Bay of Fundy,

Canada, over Three Decades," *Endangered Species Research* 39 (2019): 159–71, https://doi.org/10.3354/esr00951.

9   M. Moore, A. Knowlton, S. Kraus, W. McLellan, and R. Bonde, "Morphometry, Gross Morphology and Available Histopathology in Northwest Atlantic Right Whale (*Eubalaena glacialis*) Mortalities (1970 to 2002)," *Journal of Cetacean Research and Management* 6 (2004): 199–214.

10   M. Moore (@MklJMoore), Twitter, July 1, 2020, 10:49 p.m., https://twitter.com/MklJMoore/status/1278521082228572160.

11   "2017–2020 North Atlantic Right Whale Unusual Mortality Event," NOAA Fisheries, https://www.fisheries.noaa.gov /national/marine-life-distress/2017-2020-north-atlantic -right-whale-unusual-mortality-event.

12   M. J. Moore, G. H. Mitchell, T. K. Rowles, and G. Early, "Dead Cetacean? Beach, Bloat, Float, Sink," *Frontiers in Marine Science* 7, no. 333 (2020), https://doi.org/10.3389/fmars.2020.00333.

13   R. Pace, R. Williams, S. Kraus, A. Knowlton, and H. Pettis, "Cryptic Mortality in North Atlantic Right Whales," *Conservation Science and Practice* 3, no. 2 (2021): e:346, https://doi .org/10.1111/csp2.346.

14   J. van der Hoop, P. Corkeron, and M. Moore, "Entanglement Is a Costly Life-History Stage in Large Whales," *Ecology and Evolution* 7, no. 1 (2017): 92–106, https://doi.org/10.1002/ece3.2615.

15   J. van der Hoop, M. Moore, S. Barco, T. Cole, P.-Y. Daoust, A. Henry, D. McAlpine et al., "Assessment of Management to Mitigate Anthropogenic Effects on Large Whales," *Conservation Biology* 27, no. 1 (2013): 121–33, https://doi.org/10.1111/j.1523 -1739.2012.01934.x; and Supplement 1, https://conbio.online library.wiley.com/action/downloadSupplement?doi=10.1111% 2Fj.1523-1739.2012.01934.x&file=cobi1934-sup-0001-TableS1 .pdf.

16   L. E. Howle, S. D. Kraus, T. B. Werner, and D. P. Nowacek, "Simulation of the Entanglement of a North Atlantic Right Whale (*Eubalaena glacialis*) with Fixed Fishing Gear," *Marine Mammal Science* (2018), https://doi.org/10.1111/mms.12562.

17   D. Cavatorta, V. Starczak, K. Prada, and M. Moore, "Friction of Different Ropes in Right Whale Baleen: An Entanglement Model," *Journal of Cetacean Research and Management* 7 (2005): 39–42.

18   H. J. Myers and M. J. Moore, "Reducing Effort in the U.S.

American Lobster (*Homarus americanus*) Fishery to Prevent North Atlantic Right Whale (*Eubalaena glacialis*) Entanglements May Support Higher Profits and Long-Term Sustainability," *Marine Policy* 118 (2020): 104017, https://doi.org/10.1016/j.marpol.2020.104017.

19 D. Morin, G. Salvador, J. Higgins, and M. Minton, "Gear Analysis and Protocols: Overview of Preliminary Gear Analysis, 2007–2017," NOAA Fisheries, 1–16, https://archive.fisheries.noaa.gov/garfo/protected/whaletrp/trt/meetings/Weak%20Rope%20Subgroup/2007_-_2017_alwtrt_gear_update_4_18.pdf.

20 D. W. Laist, *North Atlantic Right Whales: From Hunted Leviathan to Conservation Icon* (Baltimore: Johns Hopkins University Press, 2017).

21 H. J. Myers, M. J. Moore, M. F. Baumgartner, S. W. Brillant, S. K. Katona, A. R. Knowlton, L. Morissette et al., "Ropeless Fishing to Prevent Large Whale Entanglements: Ropeless Consortium Report," *Marine Policy* 107 (2019): 103587, https://doi.org/10.1016/j.marpol.2019.103587; Ropeless Consortium, Woods Hole Oceanographic Institution, https://ropeless.org/.

22 Myers and Moore, "Reducing Effort in the U.S. American Lobster (*Homarus americanus*) Fishery."

23 M. Moore, "Rope-less Fishing in Practice Today: Ropeless Workshop," Woods Hole Oceanographic Institution, February 1, 2018, https://ropeless.org/wp-content/uploads/sites/112/2018/02/4.-Moore_Current_Ropless.pdf.

ACKNOWLEDGMENTS

1 North Atlantic Right Whale Consortium, Identification and Necropsy Databases, 2020 (New England Aquarium, Boston, MA), https://www.narwc.org/narwc-databases.html.

Index